Cotton Fiber: Physics, Chemistry and Biology

David D. Fang
Editor

Cotton Fiber: Physics, Chemistry and Biology

Springer

Editor
David D. Fang
Cotton Fiber Bioscience Research Unit
USDA-ARS, Southern Regional Research Center
New Orleans, LA, USA

ISBN 978-3-030-00870-3 ISBN 978-3-030-00871-0 (eBook)
https://doi.org/10.1007/978-3-030-00871-0

Library of Congress Control Number: 2018960762

This Springer imprint is published by the registered company Springer Nature Switzerland AG
The registered company address is: Gewerbestrasse 11, 6330 Cham, Switzerland

Preface

Cotton fiber is the most important natural fiber used in the textile industry. The physical structure and chemical compositions of cotton fibers have been extensively studied. Newer high-speed spinning instruments are being deployed around the world that demand longer, stronger, and finer fibers. Consequently, genetic improvement in fiber quality has been stressed. With improvement in fiber quality has come the realization that further fiber improvement will require a better understanding of fiber development and biology. As a consequence, cotton fiber developmental biology, genetics, and genomics have become focal points in the cotton research community. As the longest single-celled plant hair, cotton fiber has been used as an experiment model to study trichome initiation and elongation in plants. This book provides a comprehensive update on cotton fiber physics, chemistry, and biology that naturally separate the book into three sections. In the physics section, the physical structure of cotton fiber is first illustrated in great detail. Then a suite of fiber properties and their measuring methods are described. The pros and cons of each method are outlined. New methods to measure physical properties of single fiber and young developing fibers are included. In the chemistry section, the chemical compositions of cotton fibers are described in detail. This knowledge is necessary for efficient modification of cotton fibers for better and broader utilization. The advancement in cotton fiber modification using chemical and enzymatic methods opened new ways to utilize cotton fibers. In the biology section, the book first introduces the utilization of naturally occurring color cottons. Color cottons possess unique attributes such as better fire retardant ability. Advancement in understanding fiber color genetics and biochemical pathways and new utilization of color cottons are discussed. Recent technological advancements in molecular biology and genomics have enabled us to study fiber development in great depth. Many genes and quantitative trait loci related to fiber quality attributes have been identified and genetically mapped. Some of these genes and QTLs are being used in breeding. Progresses in cotton fiber improvement using breeding and biotechnology are discussed in the last chapter. This book serves as a reference for researchers, students, processors, and regulators who either conduct research in cotton fiber improvement or utilize cotton fibers.

I greatly appreciate all the authors who contributed excellent work to this book. I also thank my employer, USDA-ARS Southern Regional Research Center (SRRC), where most contributors are employed. Cotton fiber research and utilization has been a major research component at SRRC since its establishment in 1939. Many research results described in this book are from SRRC.

New Orleans, LA, USA David D. Fang

Contents

Contributors

Noureddine Abidi Fiber and Biopolymer Research Institute, Department of Plant and Soil Science, Texas Tech University, Lubbock, TX, USA

B. Todd Campbell USDA-ARS Coastal Plains Soil, Water, and Plant Research Center, Florence, SC, USA

Christopher D. Delhom Cotton Structure and Quality Research Unit, USDA-ARS, Southern Regional Research Center, New Orleans, LA, USA

Jane K. Dever Texas A&M AgriLife Research, Lubbock, TX, USA

Xiongming Du Institute of Cotton Research, Chinese Academy of Agricultural Sciences (ICR, CAAS)/State Key Laboratory of Cotton Biology, Anyang, Henan, China

David D. Fang Cotton Fiber Bioscience Research Unit, USDA-ARS, Southern Regional Research Center, New Orleans, LA, USA

Alfred D. French Cotton Structure and Quality Research Unit, USDA-ARS, Southern Regional Research Center, New Orleans, LA, USA

Wenfang Gong Institute of Cotton Research, Chinese Academy of Agricultural Sciences (ICR, CAAS)/State Key Laboratory of Cotton Biology, Anyang, Henan, China

Kari L. Hugie USDA-ARS Coastal Plains Soil, Water, and Plant Research Center, Florence, SC, USA

Yinhua Jia Institute of Cotton Research, Chinese Academy of Agricultural Sciences (ICR, CAAS)/State Key Laboratory of Cotton Biology, Anyang, Henan, China

Carol M. Kelly Texas A&M AgriLife Research, Lubbock, TX, USA

Brendan Kelly Texas Tech University, Lubbock, TX, USA

Texas A&M Agrilife Research, Lubbock, TX, USA

Hee Jin Kim Cotton Fiber Bioscience Research Unit, USDA-ARS, Southern Regional Research Center, New Orleans, LA, USA

Yongliang Liu Cotton Structure and Quality Research Unit, USDA-ARS, Southern Regional Research Center, New Orleans, LA, USA

Vikki Martin Cotton Incorporated, Cary, NC, USA

Marina Naoumkina Cotton Fiber Bioscience Research Unit, USDA-ARS, Southern Regional Research Center, New Orleans, LA, USA

Zhaoe Pan Institute of Cotton Research, Chinese Academy of Agricultural Sciences (ICR, CAAS)/State Key Laboratory of Cotton Biology, Anyang, Henan, China

About the Editor

David D. Fang is a Supervisory Research Geneticist and Research Leader of the Cotton Fiber Bioscience Research Unit, USDA-ARS Southern Regional Research Center in New Orleans, Louisiana, USA. He leads a team to study cotton fiber development using biochemical, genetic, molecular, and genomic methods and technologies. His research interests focus on two aspects: (1) identification of superior fiber alleles and quantitative trait loci based on the analysis of MAGIC populations, and use them in breeding with the assistance of DNA markers; (2) elucidation of the molecular mechanisms of fiber development especially elongation and secondary cell wall thickening through comparative analyses of fiber mutants and their respective wild types. Prior to the current position, he was the Molecular Cotton Breeder, and Director of Molecular Cotton Breeding at Delta and Pine Land Company. He obtained his Ph.D. in 1990 from Huazhong Agricultural University, China.

Chapter 1
General Description of Cotton

David D. Fang

Throughout the world, cotton fiber is the most widely used plant-produced fiber for apparel, home furnishings, and industrial products. In 2016, about 106.5 million bales (218 kg or 480 pounds per bale) of cotton fiber were produced from more than 50 countries around the world. The economic value of the worldwide raw cotton fiber is estimated at $35 billion annually. India, China, the United States, Pakistan, and Brazil account for over 75% of world cotton production (www.cottoninc.com). Of the entire world production, about 36 million bales were destined to the export market with the United States being the largest exporter and Vietnam as the largest importer in 2016.

Naturally, a cotton plant grows as a perennial in tropical and subtropical regions, often reaching the size of a small tree. However, for commercial production of raw fibers, most if not all cotton cultivars are grown as annuals, i.e., the crop is harvested in the same year of planting. An "annual" cotton is not a true annual because the death of a plant is not a natural consequence of seed ripening, rather due to application of chemicals or mechanical destruction. A cotton plant can be maintained indefinitely under a warm environment such as a glasshouse. Indeed, cotton germplasm repositories in several countries use glasshouses to maintain live cotton plants (Percy et al. 2014).

Cotton belongs to the family Malvaceae, the tribe Gossypieae, and the genus *Gossypium* (Wendel and Grover 2015). The cotton genus (*Gossypium* L.) consists of about 45 diploid species (2n = 2x = 26) classified as 8 genome groups (A–G and K) and 6 allotetraploid (AD) species (2n = 4x = 52) (Fryxell 1992; Wendel and Grover 2015). The haploid genome size of diploid species ranges from 885 Mbp of a D-genome species to 2570 Mbp of a K-genome species. A tetraploid species has

D. D. Fang (✉)
Cotton Fiber Bioscience Research Unit, USDA-ARS, Southern Regional Research Center, New Orleans, LA, USA
e-mail: david.fang@ars.usda.gov

D. D. Fang (ed.), *Cotton Fiber: Physics, Chemistry and Biology*,
https://doi.org/10.1007/978-3-030-00871-0_1

a genome size of about 2400 Mbp (Hendrix and Stewart 2005; Zhang et al. 2015). A preponderance of evidence has demonstrated that the six tetraploid cotton species (*G. barbadense*, *G. darwinii*, *G. ekmanianum*, *G. hirsutum*, *G. mustelinum*, and *G. tomentosum*), which are entirely New World distribution, originated from a single hybridization event between an A-genome species (either *G. herbaceum* or *G. arboreum*) and a D-genome species (possibly *G. raimondii*) 1–2 million years ago (Endrizzi et al. 1985; Paterson et al. 2012; Wendel and Cronn 2003; Wendel and Grover 2015; Zhang et al. 2015). Four species, i.e., *G. arboreum*, *G. barbadense*, *G. herbaceum*, and *G. hirsutum*, are cultivated for their ability to produce high fiber yield.

G. hirsutum (Fig. 1.1), native to Mexico and Central America, was introduced into the United States as early as the sixteenth century shortly after Columbus' discovery of the Americas (Beckert 2014). Tremendous efforts in introduction, selection, and breeding significantly improved *G. hirsutum* plants to better adapt to commercial production under new environments in subsequent centuries (Fig. 1.2). Eli Whitney's invention of a saw gin to mechanically separate cotton fiber from seeds in 1793 greatly helped the expansion of cotton production in the United States and the spread of American cotton varieties to other countries (Lee and Fang 2015). Of the four commercially cultivated species, *G. hirsutum* commonly known as upland cotton or American upland cotton is grown on the most acres and accounts for over 90% of the world's raw cotton fiber production. In general, upland cottons have fiber length ranging from 20 to 32 mm, micronaire value falling between 3.5 and 5.5, and bundle fiber strength between 27 and 32 g/tex.

G. barbadense (Fig. 1.3) originated in South America and has a wide range of distribution across the continent. This species includes commercial varieties commonly known as Egyptian, Sea Island, Pima (also called American Pima), American

Fig. 1.1 *G. hirsutum* plant (**a**) and boll (**b**) (courtesy of Doug Hinchliffe)

Fig. 1.2 *Gossypium hirsutum*. (**a**) Landrace, (**b**) cultivar (courtesy of James Frelichowski)

Fig. 1.3 *G. barbadense* (courtesy of James Frelichowski)

Egyptian, and extra-long staple. This species provides about 3–5% of the world's cotton production. *G. barbadense* varieties are mainly grown in Egypt, Sudan, and the western United States. The fiber of *G. barbadense* is longer (>33 mm), stronger (>32 g/tex), and finer (micronaire <4.0) than that of *G. hirsutum*. *G. barbadense* fibers are mainly used to produce high-quality apparel that can command a premium price. Although *G. barbadense* has better fiber, its low yield and poor adaptability to variable environments limit its cultivation. Since the beginning of the twentieth century, a lot of breeding efforts in the United States have been dedicated to introgression of *G. barbadense* fiber traits into *G. hirsutum* varieties. Many germplasm resources including commercially successful Acala-type varieties with variable levels of *G. barbadense* introgression have been developed (Smith et al. 1999).

G. arboreum and *G. herbaceum* are known as Asiatic or old world cottons (Figs. 1.4 and 1.5). They are also called "desi" cottons. These two diploid species have been cultivated by mankind for thousands of years (Lee and Fang 2015).

Fig. 1.4 *G. arboreum* plant (**a**) and boll (**b**) (courtesy of Doug Hinchliffe)

Fig. 1.5 *G. herbaceum* plant (**a**) and boll (**b**) (courtesy of Doug Hinchliffe)

Their fibers are short (<20 mm), coarse (micronaire >5.0), and weak (<22 g/tex) (Reddy and Reddy 2011), and yield is low. However, desi cottons have favorable traits such as resistance or immunity to leaf curl virus (a destructive disease affecting cottons in India and Pakistan) (Nazeer et al. 2014), blue disease (a virus disease prevalent in South American countries) (Fang et al. 2010), and bunchy top disease (another virus disease in Australia) (Ellis et al. 2016). In addition, desi cottons are drought tolerant which makes these two diploid species suitable to the arid sub-Indian continent. Desi cottons account for less than 2% of the world's cotton production and are mainly cultivated in India and Pakistan.

Cotton originated in the tropics, and the plant becomes inactive at temperatures below 16 °C. Cotton plants need about 160 days above 16 °C to produce a crop (Snider and Oosterhuis 2015; Waddle 1984). Planting time for cotton varies by locality. Planting recommendations are generally based on soil temperature greater than 16 °C at a certain planting depth and favorable air temperature forecasts for the next 3–7 days after planting. In the United States, the planting season can start as early as February in the lower Rio Grande valley in Texas and continue as late as May in the southeast regions. Soil temperature greatly affects seed germination. Seedlings emerge from the soil within a week after planting. Flower buds or squares are visible near the top part of the plant about 5–6 weeks after seedling emergence. Blossoms appear in another 3–4 weeks. The time interval from the day of anthesis (flowering) to open boll ready for harvesting is about 50–80 days depending on genotypes and environments.

Cotton fibers are unicellular trichomes or plant hairs that differentiate from epidermal cells of developing cotton seeds. Cotton fiber development occurs in a temporally ordered series of developmental stages and divides into four distinctive yet overlapping stages: initiation, elongation, secondary cell wall (SCW) biosynthesis, and maturation (Haigler et al. 2012; Lee et al. 2007). Visible signs of fiber initiation are first evident 1 or 2 days before anthesis; therefore, developmental events are staged by the number of days post-anthesis (DPA). Over a 2–3-week period, fiber cells elongate up to 25–40 mm, making them among the longest cells in the plant kingdom (Kim and Triplett 2001). During the elongation stage, only a thin (0.1–0.2 μm) primary cell wall with a waxy cuticle surrounds each fiber cell. Depending on genotypes, there are 10,000 to 20,000 fibers per seed (Zhang et al. 2011). In each boll (ovary), there can be over half a million synchronously elongating fiber cells that are in a sole cell type (Bowman et al. 2001).

In upland cotton, fiber cell starts to elongate as early as on the first day of anthesis and continues up to 20 DPA. Fiber length is largely determined at the elongation stage especially the length of elongation period as demonstrated by Avci et al. (2013) based on comparison of *G. hirsutum* and *G. barbadense* fiber length development. SCW biosynthesis begins approximately 12 to 16 DPA and continues until approximately 35 DPA or later. This stage is critical for fiber strength and maturity. The fiber SCW is deposited between the plasmalemma and the primary cell wall and is 1.5–3.0 μm thick at maturity. At or around 45 DPA, fiber development enters into maturation stage. Fiber development ceases when the fruit wall dehisces and the fibers dry upon exposure to the environment. The cytoplasmic contents of the living cell adhere to the drying cell wall.

Fiber primary cell walls (PCW) are a composite of carbohydrate polymers (cellulose, hemicellulose, and pectin) and structural wall proteins. The cellulose content of the expanding PCW is less than 15% by weight, whereas mature fibers have thickened secondary cell wall composed of nearly pure cellulose (>95%). Cellulose is a linear β-1,4-D-glucopyranose polymer that aggregates into higher-order structures called microfibrils (5–15 nm diameter and 10 μm long) (Fig. 1.6). Cellulose

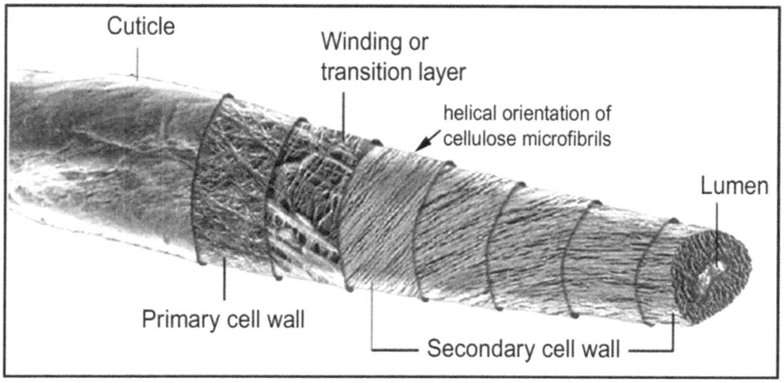

Fig. 1.6 Cotton fiber structure (courtesy of Cotton Structure and Quality Research Unit, USDA-ARS, New Orleans, LA)

microbrils (CMF) are helically arranged around the longitudinal axis of the fiber in layers. Periodically the gyre of the helix reverses direction, and a reversal is formed. There is a strong association between the orientation of CMF in the fiber secondary cell wall and fiber strength (Moharir 1998; Moharir et al. 1999; Warwicker et al. 1966). Fibers with CMF oriented with a shallow angle relative to the long axis of the fiber are stronger than fibers with larger orientation angles. The orientation of CMF in the fiber cell wall, as in other plant cells, appears to be influenced by the cytoskeleton (Seagull 1991); however, how the cytoskeleton exerts an influence on cell wall structure remains an unanswered question in plant cell biology. The degree of polymerization (number of glucose molecules per polymer) of cellulose is much larger in the secondary cell wall than in the primary cell wall (Timpa and Triplett 1993). As the cellulose degree of polymerization increases in the secondary cell wall, fiber strength increases (Timpa and Ramey 1994).

Cotton harvesting methods vary from different regions in the world. Almost the entire cotton crop in the United States is mechanically harvested; however, manual harvesting is still prevalent in many cotton-growing countries where labor cost is relatively low. There are two major types of mechanical cotton harvesters: picker and stripper. A picker harvester selectively collects seed cotton from open bolls, leaving much of the bur and other plant materials in fields. A stripper harvester collects seed cotton along with significant amount of other plant materials. Cottons harvested by a stripper tend to be dirtier and require more cleaning in the subsequent ginning process (Wanjura et al. 2015). After harvesting, seed cotton will be transported to a ginning facility. During the peak of harvesting, ginning facilities cannot immediately process all the seed cottons that were harvested. Thus, the harvested seed cottons are temporarily stored in fields in compacted modules wrapped in plastic films or covered with tarps.

Although the main function of the ginning operation is to separate fibers from seeds, the ginning process also includes conditioning (to adjust moisture), cleaning (to remove non-fiber trash), and packing into bales for transportation and marketing.

Inappropriate ginning can break fibers and greatly affects fiber quality. Upland cottons are usually ginned on saw gins, while Pima cottons are often ginned using roller gins. A cotton bale varies in dimension, volume, and weight. A typical US cotton bale weighs 218 kg (480 pounds) with dimension of 1400 mm (length) × 533 mm (width) × 736 mm (height). A cotton bale is wrapped with plastic films or cotton fabric (Wanjura et al. 2015).

Quality of upland cotton raw fibers is becoming a critical factor in cotton production in the United States. Nearly every bale of cotton produced in the United States is classed by the USDA Agricultural Marketing Service using high volume instruments (HVI) that rapidly measure fiber physical properties including length, length uniformity, strength, micronaire (MIC), and trash content. Fiber length is largely influenced by the genetic background of each cultivar, but adverse environmental conditions will reduce fiber length below a genotype's potential (Bradow and Davidonis 2000; Kelly et al. 2015; Meredith et al. 2012). Fiber length and length uniformity are important determinants for yarn strength, evenness, fineness, and spinning efficiency (Kelly et al. 2015; Thibodeaux et al. 2008). Short fibers may be generated during ginning process if the fibers are weak. Fiber strength is highly influenced by cotton genotype and may also be negatively affected by poor growing conditions (Hinchliffe et al. 2011; Zhang et al. 2017). High-speed textile processing machinery, especially rotor-spinning equipment that spins cotton fiber into yarn found in most US textile plants, puts an increased demand on higher cotton fiber strength. MIC is a measurement of the air permeability through a mass of fiber compressed to a fixed volume and is influenced by both fiber fineness and maturity. Modern high-yield cotton varieties produce high-MIC cotton (>5.0) because the yield is positively correlated to MIC value (Nichols et al. 2012). The high-MIC cotton is composed of coarse and thick fibers, and it is unfavorable to both textile manufacturers and consumers. Cotton fibers with intermediate MIC values ranging from 3.7 to 4.2 are classified as premium cotton. Buyers discount the value of high- (>5.0) and low-MIC (<3.4) cotton. In summary, the value of cotton fiber in the market, regardless of its end use, is directly related to the combination of its physical properties. These combined physical attributes of cotton fiber have a direct and significant impact on the economical return to cotton farmers and other related downstream entities. The highly mechanized production and processing of cotton products at increasingly higher speed demand that the raw fiber be as uniform, long, and strong as possible.

There are three primary products derived from cotton production: cotton lint, linters, and cottonseed. Cotton lint is long (>25 mm) fiber that can be spun into yarn. This product is used in clothing, denim, towels, and dollar bills. The lint fibers can be easily separated from seeds through the ginning process. Linters are short fibers (usually <15 mm) that are still attached to the seeds after ginning. The linter fibers are removed during the delintering process. Linters are used in plastics, paper products, films, and cosmetics. Besides length, there are many notable differences between lint and linters in physical and chemical properties (Wakelyn et al. 1998). Linters are coarser and thicker and often show pigmentation. Lint fiber cells usually initiate before or on the day of anthesis and elongate as late as 20 DPA (Avci et al.

2013). In contrast, linter fiber cells initiate at 3–4 DPA and stop elongation as early as 12 DPA. There are cotton varieties or mutants that are linter-free (e.g., *G. barbadense* varieties, N_1 and n_2 mutants) and fiberless (e.g., XZ142 *fl*) (Zhang and Pan 1991). These mutants are widely used to study the biology of fiber development (Naoumkina et al. 2016). Cottonseed is crushed into three separate products—oil, meal, and hulls (the outer covering of a seed). The oil is the cottonseed's most valuable by-product and is purified and used in cooking. The hulls are used in livestock feed, fertilizer, fuel, and packing materials. The meal is made by grinding the cottonseed and is used in livestock and poultry feed, as well as in natural fertilizers for lawns, gardens, and flower beds.

Although almost all cottons of commerce are white, naturally colored cotton fibers exist in various hues including light to dark brown, red, rust, and green, and they are found in both diploid and tetraploid species (Hinchliffe et al. 2016). Naturally colored cottons have been grown for several thousand years but almost completely disappeared in the mid-twentieth century because of the availability of inexpensive dyes, higher production of white cotton, and cotton ginners' concern of contamination to white cottons. In recent years, there is a renewed interest in growing colored cotton for better stewardship of the environment by reducing the amount of dying chemicals used to artificially color cotton fabrics. Currently, colored cottons are typically grown as a source of fiber for niche textile markets that promote the use of natural colors in textiles as an alternative to dying scoured and bleached cotton fibers. Colored cotton fibers are usually weaker, shorter, and finer and often yield lower. However, these shortcomings associated with colored cotton can be overcome through breeding if demand for naturally colored cotton fibers is increased. A recent finding that naturally colored cotton fibers confer higher flame retardancy may spark new demand for colored cotton (Hinchliffe et al. 2016).

Cotton is one of the first crops that were genetically modified using transgenic technologies. In 1996, the first transgenic cotton variety containing a Bt gene from the bacterium *Bacillus thuringiensis* was introduced to the US market. Since then, transgenic cotton has been grown in more than 15 countries. As of today, transgenic cotton accounts for more than 85% and about 60% of cotton acreages in the United States and the world, respectively (Zhang 2015). There are two major transgenic traits: Bt toxin (a protein from *Bacillus thuringiensis*) expressed in cotton varieties to protect fruit from lepidopteran insects such as boll worms and herbicide tolerance that enables easy management of weeds using herbicides such as glyphosate. Bt cotton includes a variety of genes producing different toxins developed by several companies (Luttrell et al. 2015). The first herbicide-tolerant gene to be commercialized in cotton conferred tolerance to the herbicide bromoxynil (BXN by Stoneville Pedigree Company). The BXN varieties were soon replaced by those containing genes that confer tolerance to glyphosate under the name of RoundUp Ready® (Monsanto Company). Later, RoundUp Ready Flex (Monsanto Company) and LibertyLink (Bayer CropScience) cottons were introduced. Many cotton varieties contain both Bt- and herbicide-tolerant genes. So far, no other genes controlling agronomic traits and fiber properties have been introduced into cotton via transformation with commercial success. Manipulation of fiber properties especially length and strength via biotechnology is recognized as a potential means to improve quality and develop new products.

Cotton is also on the cutting edge of genomic methods and technologies. Genome sequences of *G. arboreum*, *G. raimondii*, *G. hirsutum*, and *G. barbadense* have been published (Li et al. 2014; Paterson et al. 2012; Yuan et al. 2015; Zhang et al. 2015). Many fiber quality quantitative trait loci have been identified, and some of them are being used in breeding practices (Fang 2015; Said et al. 2013). Genes relating to fiber cell initiation (Wan et al. 2016; Wu et al. 2018; Zhu et al. 2018), elongation (Thyssen et al. 2017), and maturity (Thyssen et al. 2016) have been identified. Many more fiber genes will be identified, and the network of fiber genes regulating fiber development may be elucidated in the foreseeable future. In the subsequent chapters, the physical and chemical characteristics of cotton fiber will be described in detail. How fiber cells initiate and elongate into a 35–40-mm-long hollow tube will be illustrated. Improvement of fiber quality through conventional breeding and marker-assisted selection will be discussed.

Acknowledgment I thank Drs. Doug Hinchliffe and James Frelichowski for providing the pictures used in this manuscript. The author's research is funded by the USDA-ARS project # 6054-21000-017-00D. Mention of trade names or commercial products in this article is solely for the purpose of providing specific information and does not imply recommendation or endorsement by the US Department of Agriculture that is an equal opportunity provider and employer.

References

Avci U, Pattathil S, Singh B, Brown VL, Hahn MG, Haigler CH (2013) Cotton fiber cell walls of *Gossypium hirsutum* and *Gossypium barbadense* have differences related to loosely-bound xyloglucan. PLoS One 8:e56315

Beckert S (2014) Empire of cotton: a global history. Alfred A. Knopf, New York

Bowman DT, Van Esbroeck GA, Van't Hof J, Jividen GM (2001) Ovule fiber cell numbers in modern upland cottons. J Cotton Sci 5:81–83

Bradow JM, Davidonis GH (2000) Quantitation of fiber quality and the cotton production-processing interface: a physiologist's perspective. J Cotton Sci 4:34–64

Ellis MH, Stiller WN, Phongkham T, Tate WA, Gillespie VJ, Gapare WJ, Zhu Q-H, Llewellyn DJ et al (2016) Molecular mapping of bunchy top disease resistance in *Gossypium hirsutum* L. Euphytica 210:135–142

Endrizzi JE, Turcotte EL, Kohel RJ (1985) Genetics, cytology, and evolution of *Gossypium*. In: Caspari EW, John GS (eds) Advances in genetics. Academic, New York, pp 271–375

Fang DD (2015) Molecular breeding. In: Fang DD, Percy RG (eds) Cotton, 2nd edn. ASA-CSSA-CSSA, Madison, pp 255–288

Fang DD, Xiao J, Canci PC, Cantrell RG (2010) A new SNP haplotype associated with blue disease resistance gene in cotton (*Gossypium hirsutum* L.). Theor Appl Genet 120:943–953

Fryxell PA (1992) A revised taxonomic interpretation of *Gossypium* L. (*Malvaceae*). Rheedea 2:108–165

Haigler CH, Betancur L, Stiff MR, Tuttle JR (2012) Cotton fiber: a powerful single-cell model for cell wall and cellulose research. Front Plant Sci 3:104

Hendrix B, Stewart JM (2005) Estimation of the nuclear DNA content of gossypium species. Ann Bot 95:789–797

Hinchliffe DJ, Condon BD, Thyssen G, Naoumkina M, Madison CA, Reynolds M, Delhom CD, Fang DD et al (2016) The GhTT2_A07 gene is linked to the brown colour and natural flame retardancy phenotypes of Lc1 cotton (*Gossypium hirsutum* L.) fibres. J Exp Bot 67:5461–5471
Hinchliffe DJ, Meredith WR, Delhom CD, Thibodeaux DP, Fang DD (2011) Elevated growing degree days influence transition stage timing during cotton (*Gossypium hirsutum* L.) fiber development and result in increased fiber strength. Crop Sci 51:1683–1692
Kelly B, Abidi N, Ethridge D, Hequet EF (2015) Fiber to fabric. In: Fang DD, Percy RG (eds) Cotton, 2nd edn. ASA-CSSA-SSSA, Madison, pp 665–744
Kim HJ, Triplett BA (2001) Cotton fiber growth *in planta* and *in vitro*. Models for plant cell elongation and cell wall biogenesis. Plant Physiol 127:1361–1366
Lee JA, Fang DD (2015) Cotton as a world crop. In: Fang DD, Percy RG (eds) Cotton, 2nd edn. ASA-CSSA-SSSA, Madison, pp 1–24
Lee JJ, Woodward AW, Chen ZJ (2007) Gene expression changes and early events in cotton fibre development. Ann Bot 100:1391–1401
Li F, Fan G, Wang K, Sun F, Yuan Y, Song G, Li Q, Ma Z et al (2014) Genome sequence of the cultivated cotton *Gossypium arboreum*. Nat Genet 46:567–572
Luttrell RG, Teague TG, Brewer MJ (2015) Cotton insect Pest management. In: Fang DD, Percy RG (eds) Cotton, 2nd edn. ASA-SSSA-CSSA, Madison, pp 509–546
Meredith WR, Boykin DL, Bourland FM, Caldwell WD, Campbell T, Gannaway JR, Glass K, Jones AP et al (2012) Genotype X environment interactions over seven years for yield, yield components, fiber quality, and gossypol traits in the regional high quality tests. J Cotton Sci 16:160–169
Moharir AV (1998) True-spiral angle in diploid and tetraploid native cotton fibers grown at different locations. J Appl Polym Sci 70:303–310
Moharir AV, Van Langenhove L, Van Nimmen E, Louwarie J, Kiekens P (1999) Stability of x-ray cellulose crystallite orientation parameters in native cotton with change of location and year of growth. J Appl Polym Sci 72:269–276
Naoumkina M, Thyssen GN, Fang DD, Hinchliffe DJ, Florane CB, Jenkins JN (2016) Small RNA sequencing and degradome analysis of developing fibers of short fiber mutants Ligon-lintles-1 (Li 1) and -2 (Li 2) revealed a role for miRNAs and their targets in cotton fiber elongation. BMC Genomics 17:360
Nazeer W, Tipu AL, Ahmad S, Mahmood K, Mahmood A, Zhou B (2014) Evaluation of cotton leaf curl virus resistance in BC1, BC2, and BC3 progenies from an interspecific cross between *Gossypium arboreum* and *Gossypium hirsutum*. PLoS One 9:e111861
Nichols N, Martin V, Devine J, Li H, Jones D, Hake K (2012) Variety performance: a critical issue for cotton competitiveness. Cotton Incorporated, Raliegh
Paterson AH, Wendel JF, Gundlach H, Guo H, Jenkins J et al (2012) Repeated polyploidization of *Gossypium* genomes and the evolution of spinnable cotton fibres. Nature 492:423–427
Percy RG, Frelichowski JE, Arnold M, Campbell BT, Dever J, Fang DD, Hinze LL, Main D et al (2014) The U.S. National Cotton Germplasm Collection—its contents, preservation, characterization, and evaluation. In: Abdurakhmonov I (ed) World cotton germplasm resources. InTech., Rijeka, pp 167–201
Reddy CVCM, Reddy YR (2011) Genetic parameters for yield and fiber quality traits in desi cotton (*Gossypium arboreum* L.). J Cotton Res Dev 25:168–170
Said JI, Lin Z, Zhang X, Song M, Zhang J (2013) A comprehensive meta QTL analysis for fiber quality, yield, yield related and morphological traits, drought tolerance, and disease resistance in tetraploid cotton. BMC Genomics 14:776
Seagull RW (1991) Role of the cytoskeletal elements in organized wall microfibril deposition. Marcel Dekker, Inc, New York, pp 143–163
Smith CW, Moser HS, Cantrell RG, Oakley SR (1999) History of cultivar development in the United States. In: Wayne Smith C, Tom Cothren J (eds) Cotton: origin, history, technology, and production. Wiley, New York, pp 99–171
Snider JL, Oosterhuis DM (2015) Physiology. In: Fang DD, Percy RG (eds) Cotton, 2nd edn. ASA-SSSA-CSSA, Madison, pp 339–400

Thibodeaux D, Senter H, Knowlton JL, McAlister D, Cui X (2008) The impact of short fiber content on the quality of cotton ring spun yarn. J Cotton Sci 12:368–377
Thyssen GN, Fang DD, Turley RB, Florane CB, Li P, Mattison CP, Naoumkina M (2017) A Gly65Val substitution in an actin, GhACT_LI1, disrupts cell polarity and F-actin organization resulting in dwarf, lintless cotton plants. Plant J 90:111–121
Thyssen GN, Fang DD, Zeng L, Song X, Delhom CD, Condon TL, Li P, Kim HJ (2016) The immature fiber mutant phenotype of cotton (*Gossypium hirsutum*) is linked to a 22-bp frame-shift deletion in a mitochondria targeted pentatricopeptide repeat gene. G3: Genes Genomes Genetics 6:1627–1633
Timpa JD, Ramey H (1994) Relationship between cotton fiber strength and cellulose molecular weight distribution: HVI calibration standards. Text Res J 64:557–562
Timpa JD, Triplett BA (1993) Analysis of cell-wall polymers during cotton fiber development. Planta 189:101–108
Waddle BA (1984) Crop growing practice. In: Kohel RJ, Lewis CF (eds) Cotton. ASA-CSSA-SSSA, Madison, pp 233–263
Wakelyn PJ, Bertoniere NR, French AD (1998) Cotton fibers. In: Lewin M, Pearce EM (eds) Handbook of fiber chemistry. Marcel Dekker, Inc, New York, pp 577–724
Wan Q, Guan X, Yang N, Wu H, Pan M, Liu B, Fang L, Yang S et al (2016) Small interfering RNAs from bidirectional transcripts of GhMML3_A12 regulate cotton fiber development. New Phytol 210:1298–1310
Wanjura JD, Barnes EM, Kelley MS, Boman RK (2015) Harvesting. In: Fang DD, Percy RG (eds) Cotton, 2nd edn. ASA-SSSA-CSSA, Madison, pp 571–608
Warwicker JO, Jeffries R, Colbran RL, Robinson RN (1966) A review of the literature on the effect of caustic soda and other swelling agents on the fine structure of cotton. Pamphlet no 93. Shirley Institute, Manchester
Wendel JF, Cronn RC (2003) Polyploidy and the evolutionary history of cotton. Adv Agron 78:139–186
Wendel JF, Grover CE (2015) Taxonomy and evolution of the cotton genus, Gossypium. In: Fang DD, Percy RG (eds) Cotton, 2nd edn. ASA-SSSA-CSSA, Madison, pp 25–44
Wu H, Tian Y, Wan Q, Fang L, Guan X, Chen J, Hu Y, Ye W et al (2018) Genetics and evolution of MIXTA genes regulating cotton lint fiber development. New Phytol 217:883–895
Yuan D, Tang Z, Wang M, Gao W, Tu L, Jin X, Chen L, He Y et al (2015) The genome sequence of Sea-Island cotton (*Gossypium barbadense*) provides insights into the allopolyploidization and development of superior spinnable fibres. Sci Rep 5:17662
Zhang J (2015) Transgenic cotton breeding. In: Fang DD, Percy RG (eds) Cotton, 2nd edn. ASA-CSSA-SSSA, Madison, pp 229–254
Zhang M, Zheng X, Song S, Zeng Q, Hou L, Li D, Zhao J, Wei Y et al (2011) Spatiotemporal manipulation of auxin biosynthesis in cotton ovule epidermal cells enhances fiber yield and quality. Nat Biotechnol 29:453–458
Zhang T, Hu Y, Jiang W, Fang L, Guan X, Chen J, Zhang J, Saski CA et al (2015) Sequencing of allotetraploid cotton (*Gossypium hirsutum* L. acc. TM-1) provides a resource for fiber improvement. Nature Biotechnol 33:531–537
Zhang TZ, Pan JJ (1991) Genetic analysis of a fuzzless-lintless mutant in *Gossypium hirsutum* L. Jiangsu J Agr Sci 7:13–16
Zhang Z, Ge Q, Liu A, Li J, Gong J, Shang H, Shi Y, Chen T et al (2017) Construction of a high-density genetic map and its application to QTL identification for fiber strength in upland cotton. Crop Sci 57:774–788
Zhu QH, Yuan Y, Stiller W, Jia Y, Wang P, Pan Z, Du X, Llewellyn D et al (2018) Genetic dissection of the fuzzless seed trait in *Gossypium barbadense*. J Exp Bot 69(5):997–1009. https://doi.org/10.1093/jxb/erx1459

Chapter 2
Cotton Fiber Structure

Alfred D. French and Hee Jin Kim

2.1 Introduction

Cotton is the most important natural fiber used in textiles, and it also has other uses such as being a component of high-quality paper. Because of its importance, cotton has received a great deal of study. Still, at the time of this writing, much remains to be learned about many of the details of the cotton fiber structure. These structural details must become known to understand the relations between the structure and performance properties of the fiber. That is a prerequisite for knowledge-based improvements.

A reason for failure to learn the entire story about cotton fibers is that they are very small yet have a variable and complex organization. As seed hairs, they are complete cells that undergo a multistage biosynthesis (Chap. 7). Unlike the trunk of a tree, the cotton fiber develops its outer perimeter first and then grows inward toward the lumen at the center of the fiber. The biosynthetic tissues are themselves synthesized within the fibers, but at the end of the fiber development, they have diminished to insignificance as a fraction by weight. During the 45-day or somewhat longer development of the *Gossypium hirsutum* fiber, numerous types of structures must be constructed. Various aspects of the fiber will reflect the influence of the environment during this development. Temperature, sunlight, nutrients in the soil, and especially water are keys to the characteristics of the final product. A primary variant is the amount of secondary wall cellulose within a given fiber. That degree of

A. D. French (✉)
Cotton Structure and Quality Research Unit, USDA-ARS, Southern Regional Research Center, New Orleans, LA, USA
e-mail: al.french@ars.usda.gov

H. J. Kim
Cotton Fiber Bioscience Research Unit, USDA-ARS, Southern Regional Research Center, New Orleans, LA, USA

D. D. Fang (ed.), *Cotton Fiber: Physics, Chemistry and Biology*,
https://doi.org/10.1007/978-3-030-00871-0_2

thickening, or maturity, of the fiber is a primary quality parameter (measured as maturity ratio, or by inference, micronaire; see Chap. 3) as it is a major factor in processing and one that impacts the response of the fiber to dye. Even for fibers on the same seed, different amounts of nutrients will be available, and the crowding of fibers in the developing boll will result in different growth environments.

Cotton has the distinction of being, with only mechanical cleaning, quite pure cellulose, as much as 95%. Cellulose, the sugar of little cells, is a polymer of as many as 20,000 glucose residues linked β-1→4. Figure 2.1 summarizes many of the widely accepted, if incompletely understood, components and properties of the fiber. Later in this chapter, a revision is proposed. This figure is a montage, and the individual segments are not presented to a constant scale. The dimensions are for a typical upland (*G. hirsutum*) cotton fiber; other cotton species may differ considerably in fiber diameter and length. Chapter 4 discusses more of the composition of the cuticle and primary wall; it suffices for this chapter to state that the cuticle and primary wall are the locations of waxes, pectins, and other polysaccharides, as well as various sugars and metals. Substantial amounts of these components are often removed during processing, leaving the cellulose component behind. The winding layer (Fig. 2.2) is associated more with the secondary cell wall fibers; little is known about it.

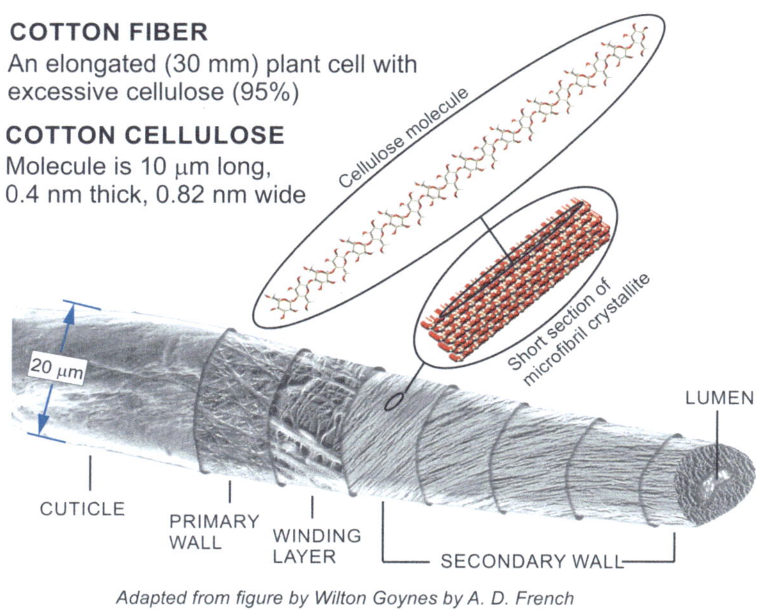

Fig. 2.1 Montage of electron micrographs, not to scale, selected and placed to resemble the different layers that compose the cotton fiber that is modified from the original figure described in Goynes (2005). In particular, the progressive change in the orientation of the microfibrils to the fiber axis as the fiber is penetrated may not be correct; see the section below on synchrotron diffraction of single fibers

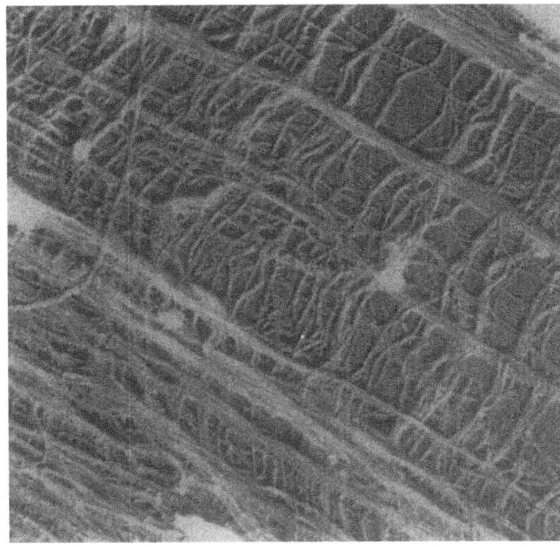

Fig. 2.2 Transmission electron micrograph of cotton fiber winding layer from the Southern Regional Research Center's archives. Previously published in Wakelyn et al. (2007)

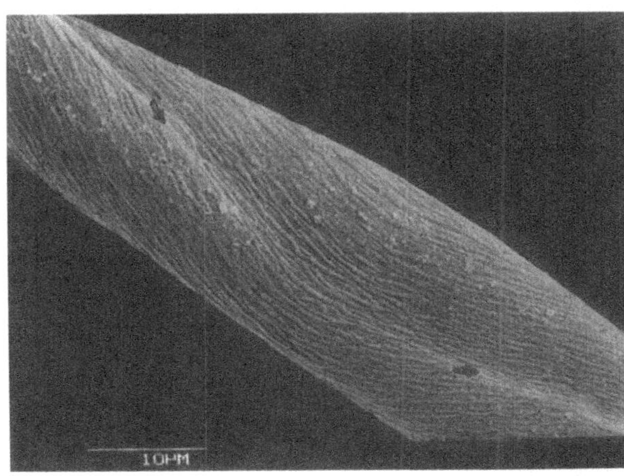

Fig. 2.3 Cotton fiber section showing surface effects with black arrows that indicate the internal convolution or reversal of crystallite orientation. Previously published in Wakelyn et al. (2007)

Not indicated in Fig. 2.1 are the convolutions that develop as the fiber dries. Fibers taken from freshly picked unopened bolls and never dried are generally rounder in cross section (Rajasekaran et al. 2006). When allowed to dry out, they flatten, putting various stresses that may be associated with the convolutions. These convolutions appear to be reversals in the underlying orientations of the microfibrils. According to Han et al. (1998), there are about ten of these convolutions per mm. These convolutions, or reversals, Fig. 2.3, are thought to engender weak points along the fiber. Because of those weak points and the fact that the molecule is much shorter than the fiber, the breaking strength and modulus of cotton fibers are much less than the theoretical values for a crystal of cellulose (Santiago Cintrón et al. 2011; Dri et al. 2013).

The hierarchy of the dimensional scales is summarized as follows: with a 0.5 nm length per glucose residue, cellulose molecules are of the magnitude of 10 μm long, or roughly as long as a fiber is wide. A 30 mm fiber would be the length of 3000 cellulose molecules end to end, and the distance between convolutions would be similar to the length of 10 molecules. Evidence based on leveling-off degree of polymerization studies indicates that the crystallites might be 175 glucose units long (~8.5 nm) (Krässig and Käppner 1961); a single molecule could participate in some 100 different crystallites. Along these same lines, an individual crystallite of cotton cellulose might host some 50 molecules as they pass through; the crystallite cross section might be some 5 nm wide.[1] In a 10 μm diameter fiber, there could be some two million crystallites in any given cross section, assuming for an order of magnitude estimate that there is no lumen and the fiber is circular in cross section. If there are 3000 molecules end to end, and each passes through 100 crystallites, there could be ($300{,}000 \times 2{,}000{,}000 = 6 \times 10^{11}$) crystallites in a fiber.

The remainder of this chapter will focus on the secondary cell wall, ignoring the other components of the fiber. The secondary wall is treated here as if it is composed of only cellulose; the other tissues in the fiber are, for all practical purposes, the locations of the other fiber components (see Chaps. 4 and 5). There is some evidence that the secondary wall is built in layers that are governed by the diurnal cycle with the accompanying temperature fluctuations (Haigler et al. 1991). Although the other tissues in the cell wall are ignored herein, they should not be forgotten. For example, when two fibers touch each other, it would be their cuticles or primary walls that are in contact, not the secondary wall material. Although examples will be taken from work on cotton, more definitive information on the structure of apparently similar cellulose has been gained from samples from other sources, and they will be discussed as well. Furthermore, we will use computerized models to indicate the interplay of structure and diffraction evidence.

2.2 Structure of the Secondary Wall

2.2.1 Diffraction

In the first three decades after the beginning of the twentieth century, several advances in science took place that established the basic ideas about the chemical and physical structures of cellulose and other polymers. Several sections of Zugenmaier's book (2008) provide more history than can be given here. Of these advances, none played a more important role than diffraction. Diffraction analyses continue to play a major role in studies of cellulose, both in terms of determination

[1] The values of crystallite width and number of chains are based on measurements of diffraction pattern peak widths at the end of this chapter. The values were interpreted to give a "diamond-shaped" model. Krassig and Kappner's length of 85 nm is longer than the 21.2 nm value indicated by the diffraction pattern in Fig. 2.10.

of exact details of structure and as a technique for more or less routine monitoring of series of samples that have been treated one way or another to learn what changes have taken place. For this reason, many materials science research papers on cellulose include diffraction studies. Mainly, diffraction studies use X-ray radiation, but neutron beams (from a nuclear reactor) and electron beams (in an electron microscope) have also been extremely helpful in working out the mysteries of cellulose structure.

For crystals as large as 200 μm, crystallography based on a single crystal becomes the ultimate analytical technique. The individual atoms can be located with four or more significant figures, and the determined number of electrons indicates the type of atom. This enables definition of the covalent structure and geometry (bond lengths and angles as well as molecular shape) to useful accuracy. One particular reason for high accuracy is to test various computer modeling methods that can be used to probe structural aspects of cellulose for which experimental data is of substantially lower quality.

Charge density studies based on extra high-quality X-ray data can also visualize the hydrogen bonding structure and even reflect van der Waals interactions (Stevens et al. 2010). Many small organic molecules, such as di- and oligosaccharides of glucose, can form sufficiently large crystals and can be studied by the conventional, single-crystal techniques. One very relevant structure is that of β-cellobiose (Kalenius et al. 2008). The charge density, also called atoms-in-molecules or electron density gradient analysis approach (Bader 1990), often involves a conventional X-ray diffraction structure determination as well as more extensive data collection and work-up. In the case of ordinary or low-quality data, the coordinates can be used as input into a quantum mechanics (electronic structure theory) program, and the bonding can be examined based on the electron density instead of geometric criteria. Such an approach has been applied to the many van der Waals and hydrogen bonding interactions in cellulose (French et al. 2014).

Somewhat smaller crystals of importance can also be studied with conventional single-crystal methods if synchrotron radiation is used. For example, cellotetraose, the fraction of a cellulose chain that contains four glucose residues, was finally solved using such radiation (Gessler et al. 1995). Synchrotron beams are typically orders of magnitude more powerful than beams from laboratory X-ray generators. Synchrotrons can give higher-resolution diffraction patterns but can also destroy the sample.

In a typical diffraction experiment on cellulose fibers, a monochromatic beam of X-rays irradiates the sample. The main beam passes through the sample and is stopped by, for example, a piece of lead. However, based on constructive interference of waves from the excited electrons in crystalline samples, diffracted rays are emitted and collected by a recording device, such as photographic film, scintillation counter, or CCD (charge-coupled device) detector similar to a digital camera. An apparatus for such experiments could also be used for single-crystal studies. However, for many types of cellulose samples, it is not convenient to prepare bundles of fibers. Instead, samples such as nanocrystals or pulp or microcrystalline cellulose are treated as powders, and powder diffractometers are used. They are optimized to collect data rapidly in reflection mode and are more often used for routine cellulose studies. Examples from both types of systems are used herein.

The individual diffracted beams from a single crystal result in spots on the recording device. If many crystals are randomly oriented and the crystals are less than a micron in size, in other words a powder, a pattern of circular rings will be recorded. Typical fibers are an intermediate type of sample, with the spots of the single crystal or ideal fiber pattern spread over small arcs that indicate the extent of misalignment of the crystallites with the fiber axis. The rotated single crystal is replaced by large numbers of crystallites aligned with the fiber axis but randomly rotated about their individual axis that is aligned with the fiber axis. Figure 2.4 indicates the types of sample and the resulting patterns. Cotton fibers are more complex, however. It appears that the microfibrils depicted in Fig. 2.1 are equivalent to the fibers described in this paragraph, each with a rotational distribution of individual crystallites but with those crystallites aligned with the microfibril axis that, in turn, is aligned at various angles to the axis of the entire cotton fiber.

Four important kinds of information are contained in each diffraction spot or ring. Their positions are determined by the dimensions of the unit cell. The unit cell is a parallelepiped; its contents, when repeated by translations along the unit cell edges, can generate the entire crystal. Second, the intensity of a spot or ring depends on the location and kind of the atoms within the unit cell. The third data type from each spot or ring is its breadth, which can be influenced by the size of the crystal and

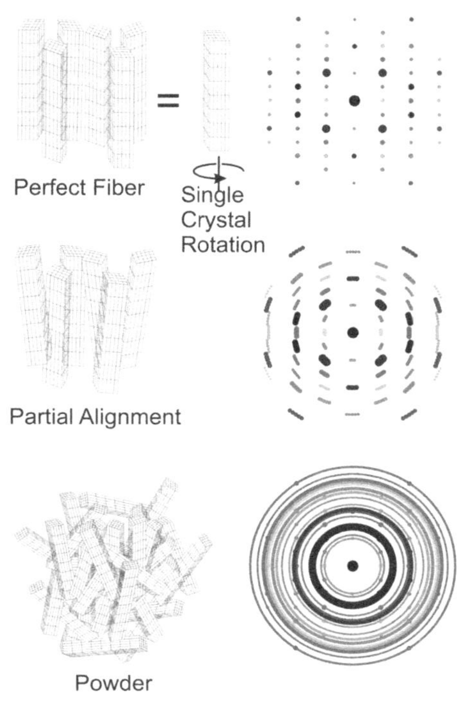

Fig. 2.4 Cartoons of crystalline samples and associated diffraction patterns collected on a two-dimension detector such as a flat film camera or a CCD 2-D area detector. The partially aligned crystals would give a pattern similar to a fiber bundle. Smooth rings in a powder pattern depend on the random orientation of many small crystals

various other factors such as disorder and thermal motion. The characteristic variation in intensity around the powder ring, or the extent of arcing of the spots on a fiber pattern, indicates the extent of nonrandom orientation of the crystallites or the degree of microfibril orientation (microfibril angle, MFA).

Diffraction patterns of many different samples of native and chemically manipulated cellulose have indicated that cellulose is a polymorphic material. In other words, cellulose can have several different forms or allomorphs (French 2014). When homogeneous derivatives of cellulose are formed, the molecules have been dissolved, and the resulting crystal structures are different because the chemically different molecules will not fit into the original lattice. However, when heterogeneous reactions are carried out, the surfaces of the crystallites are the major locations of the substitution, and the arrangements of the internal chains need not be altered. Diffraction studies are often carried out before and after chemical modification to learn the extent of changes.

2.2.2 Cellulose Crystal Structure

For the present discussion, we are concerned only with the crystal form cellulose I_β, considered to be the form of most of the crystallites in native cotton. Noncrystalline cellulose is also of interest. Diffraction studies of crystallites of tunicate cellulose (Nishyama et al. 2002) established the structure of this form of cellulose, determining the unit cell dimensions and atomic positions to a high degree of confidence. It might seem unusual that tunicate cellulose from a marine animal (*Halocynthia roretzi*) is the best model for plant cellulose, but its unusually large crystals permit the most complete structure determination. More details of the molecular structure follow, but we begin to make use of the basic information at this point.

Figure 2.5a, b shows the basic dimensions of the I_β unit cell from tunicate, although small deviations from these values are observed for cellulose from other sources that are of greater practical interest such as cotton. The monoclinic unit cell has *a*, *b*, and *c* dimensions as indicated. Characteristic of monoclinic unit cells, the angles between the *a*- and *c*-axes and between the *b*- and *c*-axes are 90°, whereas the monoclinic angle, γ, is 96.55°. Also shown in Fig. 2.5a, b are the two glucose residues of the asymmetric unit. One is centered at the corner of the unit cell, and a slightly different one is placed at the center, shifted along the *c*-axis by about 1/4 of the 10.38 Å *c*-axis length from the corner chain. This unit cell is not fixed relative to the glucose residues along the *c*-axis. Only one of the O4 atoms on each residue in Fig. 2.5b is officially part of the repeating unit. The entire cellulose crystal can be generated by repetition of the asymmetric unit by employing the symmetry operators of the $P2_1$ space group and translational symmetry along the *a*-, *b*-, and *c*-axes.

The unit cell dimensions, the crystal symmetry, and the coordinates of the atoms of the glucose residues in the unit cell are the minimum information needed to define the crystal structure. More residues of the cellulose molecules are generated by repeated application of the twofold screw-axis symmetry operator of the

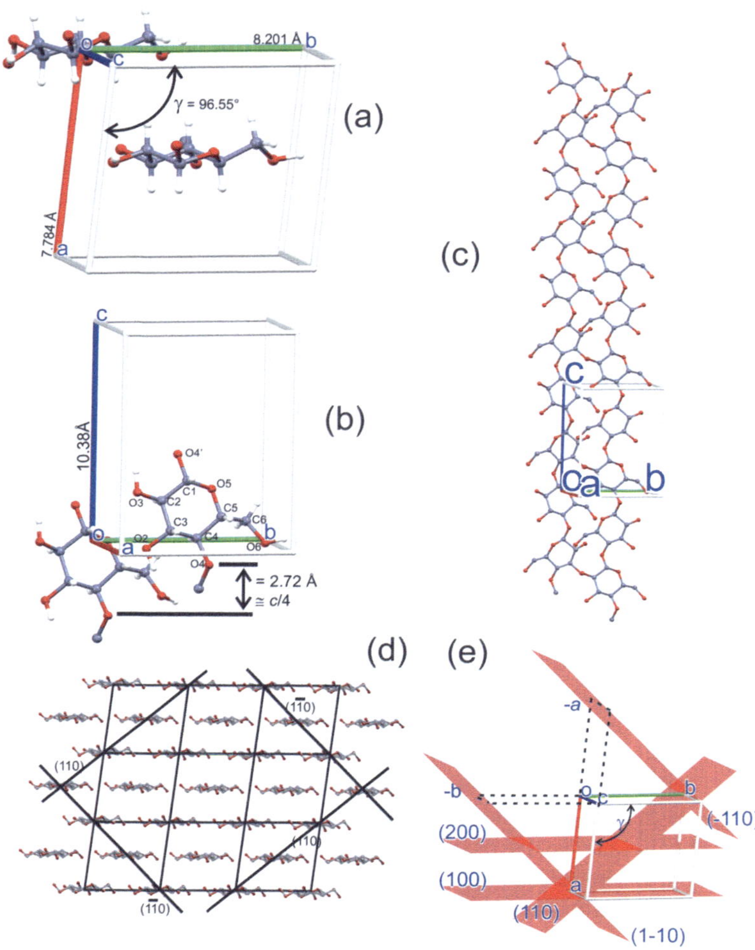

Fig. 2.5 (**a**) The asymmetric unit of the cellulose I_β unit cell, along with the unit cell dimensions. The *c*-axis repeat length, not shown, is 10.38 Å; the *α*- and *γ*-angles are 90.0 as dictated by the *P2₁* monoclinic symmetry. The visual projection is near to down the *c*-axis; the origin is designated by "o." An extra linkage oxygen atom (O–1 or O–4) is shown for each of the glucose residues. (**b**) As in (**a**), but projected nearly down the *a*-axis. The "quarter stagger" or "shift" of the central chain is indicated, as is the numbering of the atoms of the glucose residue. (**c**) The corner and center cellulose molecules passing through a unit cell. Cellodecaose fragments are shown. (**d**) A model crystal generated from the asymmetric unit, viewed exactly down the *c*-axis. The horizontal lines correspond to the (100) crystal planes; the nearly vertical lines correspond to the (010) planes. Indicated on the drawing are the (11¯0) and (110) planes. Deletion of the eight intact molecules outside these planes removes molecules that are less stabilized by their neighbors and results in a model crystal with 23 cellulose molecules. (**e**) Three-dimensional view of the important equatorial crystal planes. All of these images were created with the Mercury software (http://ccdc.cam.ac.uk/solutions/csd-system/component/mercury)

monoclinic space group *P*2$_1$. That *P*2$_1$ operator determines that for every atom with the coordinates of *x*, *y*, and *z*, there is an equivalent atom at −*x*, −*y*, and *z* + 1/2. A third residue in each chain has the coordinates of *x*, *y*, and *z* + 1. Figure 2.5c shows two eight-residue fragments of cellulose chains (cellodecaose), one at the corner and one at the center of the unit cell.

More molecules in the crystallites are created by application of translational symmetry along the edges of the unit cell. Figure 2.5d shows a small model crystal, created by such translation operations along the *a*- and *b*-axes. This model is smaller than cotton crystallites. New models can be made by starting with a big model and deleting chains to make a final product. For example, a "hexagonal" model could be made by keeping the central row of five molecules, the two rows above and below the central row with four residues each. Proceeding to the next rows further from the central row, keep three residues in each, and, finally, keep just two chains. Such a model would have 23 molecules, perhaps a bit bigger than a wood cellulose crystal (Fernandes et al. 2011).

An important concept in diffraction pattern analysis is the reciprocal lattice. The diffraction pattern is essentially a recording of the reciprocal lattice. Long distances in "reciprocal space" correspond to short distances in the unit cell. Another concept is the presence of planes in the real lattice, from which the X-rays are "reflected." These planes pass through the unit cell in such a way that they intersect with the edges at various integer fractions of the unit cell. For example, in Fig. 2.5e, the (200) plane intersects the *a*-axis at one half the distance from the origin, but it never intersects the *b*- or *c*-axes. Referring to Fig. 2.5d, it is seen that electron density is concentrated in the planes that intersect the *a*-axis at intervals of *a*/2, causing the (200) reflection to be most intense on most cellulose I_β patterns. Another important plane is (110), also in Fig. 2.5e. It intersects both the *a*- and *b*-axes at values of 1 × *a* and 1 × *b*, respectively. The (1–10) [also written (1¯10) and pronounced "one, one-bar, zero"] plane is just as important. That plane intersects the *a*-axis at +1 and the *b*-axis at −1. It is entirely equivalent to the (¯110) [or (−110)] plane. Because of the presence of adjacent unit cells in crystals, another member of the (1¯10) family of planes passes through the origin. The model with 23 chains (Fig. 2.5d) in the previous paragraph is bounded by the (100), (1¯10), and (110) planes.

All of this depends on the convention used to describe the unit cell. Two conventions are given in the International Tables for Crystallography for space group *P*2$_1$, one with the monoclinic axis as *b* (second setting) and one using *c* (first setting.) Although studies of most materials with monoclinic symmetry choose the *b*-axis, for polymers it is preferred to use the *c*-axis because of polymorphy. Thus in soda cellulose II, the unit cell is hexagonal (Whitaker et al. 1974), and convention dictates that *a* = *b*, leaving the *c*-axis unique. The *c*-axis can be unique for all space groups (all axes are unique in the triclinic space groups). It is important for readers to be aware that numerous conventions have been used and that novice users will incorrectly mix the Miller indices (the values of *h*, *k*, and *l*). Authors are encouraged to use the conventions employed herein (French 2014).

Initially it was thought that an entire cellulose molecule must fit inside the unit cell. Confusion and incorrect structural proposals resulted (Purves 1943). Subsequently it was realized that an "infinite" polymer could also be characterized by a unit cell such as shown here (Sponsler and Dore 1926). The necessary concept is that the ends of the cellulose chain, which are not identical to the other glucose residues, are such a small fraction of the mass that they do not have a significant effect on the diffraction pattern.

It is important to understand that diffraction data has an underlying basis that is different from those of either vibrational (IR or Raman) or magnetic resonance (NMR) spectroscopy. The individual peaks on a cellulose powder diffraction pattern do not generally individually represent a particular atom such as the chemical shift for the C–4 atom in the glucose ring, or a functional group, such as the stretching mode for a C=O bond. The peaks on the pattern arise from the constructive interference of radiation from coplanar arrays of atoms that are determined by the unit cell dimensions. All atoms nominally contribute to all of the peak intensities that depend directly on the crystallite size and lattice perfection. So, a diffraction pattern is not really a "spectrum," and the peaks are not "bands." Another misunderstanding in some of the literature is the incorrect claim that one of the peaks (the overlapped (1¯10) and (110) peaks, see Fig. 2.6a) represents amorphous material, while the main (200) peak represents crystalline material. All of the peaks in Fig. 2.6a are based on the coordinates of the atoms in the crystal model, all of which are interrelated by simple translations along the unit cell edges.

Fourier transform infrared (FT-IR) spectroscopy has been used to distinguish secondary cell wall from other cell wall components in a rapid and noninvasive way from other model plants. Despite different definition of crystallinity of X-ray diffraction (XRD) from infrared methods, the XRD crystallinity values determined

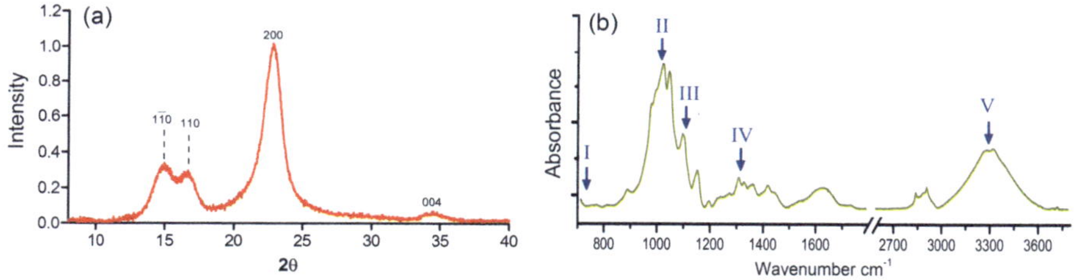

Fig. 2.6 Comparisons of XRD with IR analyses of raw *G. hirsutum* TM-1 fibers. (**a**) X-ray diffractograms; linear backgrounds were subtracted from raw data. The peaks associated with the different lattice planes of Fig. 2.5d, e are shown, as well as the (004) peak. This one-dimensional experimental pattern was taken by a powder diffractometer using Cu Kα radiation and corresponds to a trace across the rings of the 2-D powder pattern in Fig. 2.4. (**b**) Attenuated total reflectance FT-IR spectra. Five regions including I (700–760 cm^{-1}), II (950–1020 cm^{-1}), III (1100–1200 cm^{-1}), IV (1260–1340 cm^{-1}), and V (3200–3400 cm^{-1}) are contributed by cellulosic components; other peaks may be due to pectin, waxes, and water. See also Lee et al. (2015)

from cotton fibers were used as the standards for those estimated by the Fourier transform infrared (FT-IR) spectroscopy (Liu et al. 2012; Abidi et al. 2014). Recently, several FT-IR methods have been proposed for monitoring of secondary cell wall cellulose levels in a quantitative and rapid way during cotton fiber development (Abidi and Manike 2017; Kim et al. 2018). XRD and FT-IR patterns are compared from Texas Marker-1 (TM-1) fibers that are a standard upland cotton for genetic and genomic studies (Fig. 2.6a, b).

Nuclear magnetic resonance (NMR) spectroscopy has made numerous contributions to knowledge of cellulose structure. The individual atoms have unique environments, and this fact results in different chemical shifts. These shifts can be simulated with computational chemistry from different models; the model that best simulates the experimental spectrum in question can be considered the most likely. Fine details of the interactions of cellulose and water have been worked out, for example (Yang et al. 2018).

Returning to diffraction, experiments are carried out on several different forms of cellulose. For laboratory X-ray systems, the samples can be in the form of fiber bundles (spun yarns serve well), fabrics, films, or powders. The biggest difference for materials based on the cellulose polymer is that, with the exception of some selected area electron diffraction carried out in an electron microscope (Sugiyama et al. 1991), the samples all contain many crystallites. Powder X-ray diffractometers used for the routine cellulose samples operate in reflection mode, allowing large samples and relatively rapid data collection. Figure 2.7 displays an experimental electron diffraction pattern from a relatively large area of a fragment of cotton secondary wall and a simulated pattern (Nishiyama et al. 2012) from a cellulose model with 81 cellulose molecules, each 20 glucose units long. The agreement between the patterns, indicated in part by the equatorial and meridional scans, suggests that the approximate understanding inherent in the model is correct.

2.3 Cotton Structure

It comes as a surprise to many that cotton fibers would be subjects for studies of their crystal structure. The layperson would not expect something as flexible and soft-feeling to be crystalline. The catch is that cotton fibers contain very small crystals, probably fewer than 100 molecules. The diffraction patterns observed under early conditions led some crystallographers to conclude that cellulose is amorphous and that it has no crystalline structure. The International Union of Crystallographers considers a material to be crystalline if it gives relatively sharp diffraction peaks. The obvious question is "What is relatively sharp?" For our purposes, a sample will be considered to have crystalline order if its powder pattern has more than one diffraction maximum; many cellulose samples have more peaks than that as seen in Fig. 2.6a. However, no cellulose fiber is able to provide as many peaks as typical samples that are more than 100 μm in size used for single-crystal studies.

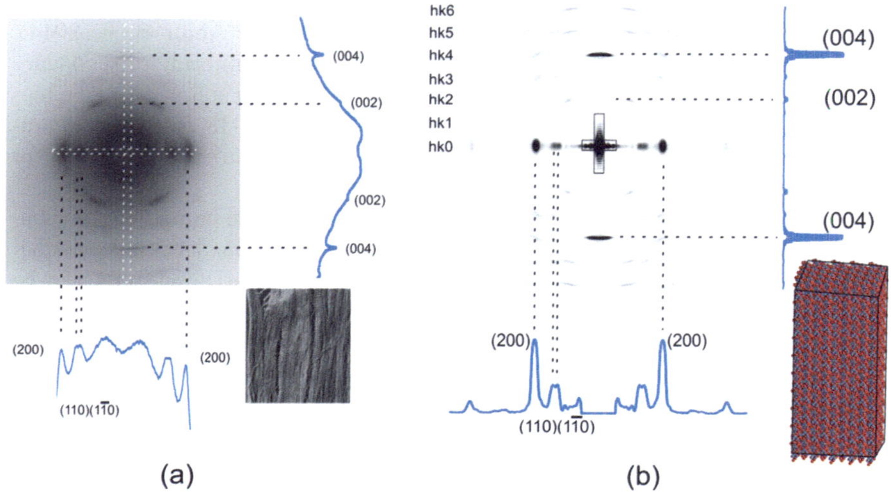

Fig. 2.7 (**a**) Experimental electron diffraction pattern for a secondary wall fragment of cotton fibers (Del Cerro 8 variety). Also shown are intensity traces based on the paths enclosed by the dashed white lines, with labels for the dominant peaks. The background halo is quite strong. A major result is that the (004) reflections are much more narrow than the (200) peaks, indicating through the Scherrer equation that the crystal is much longer along the molecule than it is perpendicular to the molecule. The intensity ratio of the (200) peak to the (11¯0) and (110) peaks is much lower than for an ideal pattern, probably because of preferred orientation of the sample. The electron diffraction intensities are not expected to be the same as X-ray intensities, either. (**b**) Calculated fiber diffraction pattern computed for the explicit 81-chain model; each chain is composed of 20 glucose residues. Again, the (004) peaks are considerably sharper than the (200) peaks. Because of the sharp boundaries of this explicit model, there is substantial small-angle scattering near the center that corresponds to the dimensions of the model crystal. Some of this scattering is visible near the more intense reflections, as well. Also shown are the general labels for reflections on the equator (hk0) and on the first (hk1) layer line, etc.

2.3.1 Interpretation of Diffraction Patterns

This discussion will start at a fairly fundamental level and build up to the type of patterns seen in routine studies such as Fig. 2.6a. The even more fundamental determinations of the atom positions in the unit cell are presented in Nishyama et al. (2002).

2.3.1.1 Electron Diffraction of Secondary Wall Fragment

Figure 2.7a is an experimental electron diffraction pattern taken from a layer of the secondary cell wall such as shown in the accompanying micrograph. Because the pattern is from electron diffraction, the relative intensities of the spots are not expected to be the same as on an X-ray pattern. However, there is no such X-ray pattern for such a portion of a fragment of a layer of cotton secondary wall from one fiber; the amount of material is too small. (Other cellulose samples such as ramie or flax can give patterns with short arcs.) Most of the microfibrils in the micrograph are reasonably parallel, resulting in short arcs on the diffraction pattern. Note that electron diffraction of cellulose is difficult to record; the electron beam can easily burn the sample before the image can be captured.

Using ImageJ software (Rasband WS (1997–2016) ImageJ, U. S. National Institutes of Health, Bethesda, Maryland, USA, https://imagej.nih.gov/ij/), the pattern was scanned in the two areas indicated by the white dashed lines. These areas divide the pattern into four quadrants, each of which is a mirror image of two others (true for monoclinic, orthorhombic, and cubic space groups). The results of these scans are plotted to the right for the vertical area and below the pattern for the horizontal area.

The calculated or simulated X-ray pattern in Fig. 2.7b is related. Indicators of the layer lines have been added. The horizontal line labeled hk0 is called the equator, and a vertical line through the center is called the meridian. Spots on the horizontal line labeled hk2 would be said to be on the second layer line. Instead of the many microfibrils in the electron diffraction pattern, the simulated sample is a crystal model that is rotated about its vertical axis.

Scans of the calculated and observed patterns are similar. The smaller areas in the boxes at the center of the pattern are excluded, however. Those areas include the extremely strong central beam and the small-angle scattering that contains information on the size of the model crystal. Not all small-angle scattering was suppressed for the horizontal scan of the (hk0) reflections. Relative intensities of the equatorial reflections (starting with the $(1\bar{1}0)$ peak and moving out from the center) as shown by the blue scan line are very similar to those found on a routine powder X-ray diffraction scan. The vertical scan shows the meridional (00l) reflections. This image resembles a precession camera diffraction pattern (or electron diffraction pattern) which enables the (00l) reflections to be observed at high intensity compared to a more typical flat plate camera. Besides the (002) and (004) reflections, there is a faint trace of a (003) reflection visible on the scan and the pattern. Further the layer lines (hk1, hk2, etc.) have hyperbolic curvature on a flat plate X-ray camera, although the very short wavelengths used for electron diffraction do not result in much curvature. The spots on this calculated pattern are arced by imposing a Gaussian distribution on them.

The electron diffraction pattern in Fig. 2.7a has a pronounced background, and the intensity of the (200) reflections is only a bit greater than the overlapping $(1\bar{1}0)$ and (110) peaks, whereas there is no background on the calculated pattern of Fig. 2.7b, and the relative intensity of the (200) peak is much greater. The peaks on the meridians of both patterns are sharper than those on the equator, reflecting the presence of more repeating units along the fiber axis than perpendicular to it. The most important difference between the simulated and experimental patterns here is the larger number of simulated diffraction spots that have been recorded.

Electron diffraction is seldom performed because once the difficult experiment is carried out, the detailed analysis of the patterns is also difficult. Fiber diffraction patterns are much more widely reported for other cellulosic materials such as the linen and ramie textile fibers, as well as the tunicate (*Halocynthia roretzi*) that has the largest crystals of the cellulose allomorph that is thought to be reasonably similar in structure to the crystals of cotton cellulose. Although the model crystal is too short, it is similar to cellulose nanocrystals (CNCs) that can be obtained from cotton fibers by treatment with 67% sulfuric acid. That treatment, however, reduces the cotton to a very fine powder; there has been little effort to orient the cotton-based CNCs to make a suitable sample for fiber diffraction. One way would be to make a composite with a stretchable, amorphous polymer; the act of stretching strips of the polymer should align the crystallites. Secondly, the microcrystals or nanocrystals could be induced to form a film by drying a slurry inside a glass vial. The vial is rolled about its long axis to align the nanocrystals during the drying process (Nishiyama et al. 1997).

2.3.1.2 Synchrotron X-Ray Diffraction of Single Fibers

Figure 2.8 shows two single-fiber diffraction patterns taken with synchrotron radiation. The very low mass of a single fiber requires a very tightly collimated beam that is very powerful. The pattern on the left was taken with a 1-μm beam in the center of the fiber as part of a series of eight patterns spaced 5 μm apart along the fiber axis.

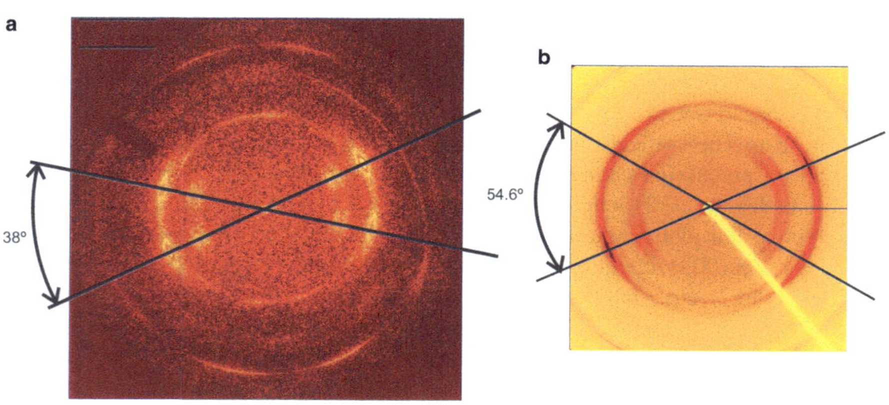

Fig. 2.8 (**a**) A diffraction pattern provided by Christian Riekel at the European Synchrotron Radiation Facility in Grenoble, France. The pattern can be understood by considering it to result from two microfibrils at a mutual angle of 38° or each 19° from the fiber axis that is approximately vertical. (**b**) A pattern from Lee Makowski at Northeastern University who collected the data at the Brookhaven National Laboratory synchrotron. Again, it is a single fiber, but the beam width was 5 μm, leading to inclusion of data from many more crystallites. Again, there are two dominant orientations of the crystallites or microfibrils, this time with a greater deviation from the fiber axis. As described further, these patterns, especially pattern a, provide evidence that contradicts the model fiber shown in Fig. 2.1, in which there is a progressive change in the microfibril orientation progressing from the outer layers of the secondary wall. The fiber patterns were colorized with XRD2DScan software (Rodríguez NA, http://www.ugr.es/~anava/xrd2dscan.htm)

Other patterns in this series had progressively smaller angles between the two layer lines.

These patterns, especially the one on the left, dispute the microfibril angle relative to the fiber axis (MFA) that is indicated in Fig. 2.1. In Fig. 2.1, the MFA undergoes a progressive change at any one location along the length of the fiber. That is accomplished by different layers of secondary wall, each with its own MFA. What these patterns demonstrate is that the MFA varies with position along the length of the fiber, but at any one position, it has a particular value. Since a series of eight patterns was taken over about 40% of the distance between convolutions, we are proposing that the X-shaped arrangement of the two equators is due to an arrangement something like the proposal illustrated in Fig. 2.9. The angle of the microfibrils to the fiber axis will varies, following the convolutions as indicated in Fig. 2.2.

In Fig. 2.9, a narrow (e.g., 1 μm) X-ray beam striking the center of the fiber would only encounter a narrow range of fibril orientations, but a random orientation of crystallites around each microfibril axis would lead to an intensity distribution along the equator similar to the horizontal trace in Fig. 2.7b. At other places between the convolution beginning and end, the MFA will be different, approaching zero at some point.

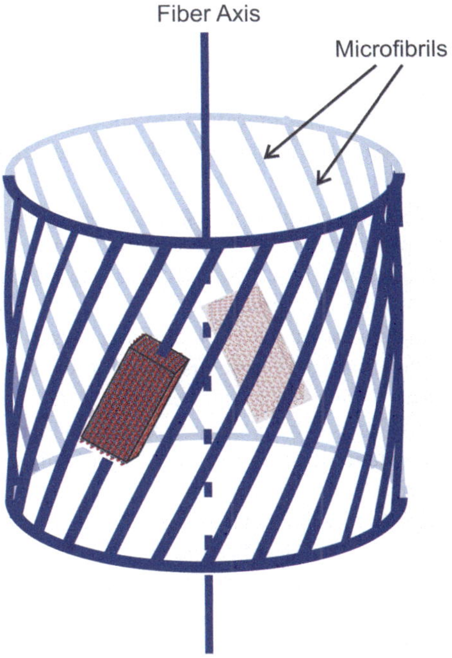

Fig. 2.9 A proposal for the arrangement of microfibrils of cotton secondary cell walls. The microfibrils are composed of crystals that are in random orientations about each of the microfibril axes. Two diffraction patterns would be produced that are similar to the pattern in Fig. 2.7b but tipped so that the meridian in each is tilted to match the microfibril axis in the front of the fiber and the back of the fiber. This is similar to drawings in Liu et al. (2016)

2.3.1.3 Synchrotron Study of Cotton Fiber Bundle

In the following discussion, all of the diffraction patterns were collected with a radiation wavelength of 1.5418 Å, the most common wavelength used on laboratory X-ray tubes. It was also used at the CAMD synchrotron.[2] The wavelength used in Fig. 2.6 was also 1.5418, from Cu Kα radiation. The wavelengths in Figs. 2.7a and 2.8 were different, and the spacings in the simulated pattern in Fig. 2.7b are proportional to $1/d_{hkl}$ as if the patterns were collected with a precession camera. Many reported diffraction patterns do not specify the radiation used, and the 1.5418 wavelength would be the first guess for a laboratory powder diffractometer. If the wavelength is different, the peaks will have different 2θ positions so it is imperative to specify the radiation. Older work reported the *d*-spacing values before the unit cells were agreed upon (e.g., Sisson 1935). One very convenient way to check patterns taken with different radiation sources (molybdenum, chromium, cobalt, and iron are also sometimes used) is to simulate the pattern, for example, with the Mercury software. A "crystal information file" (.cif) is needed for input for the different allomorphs of cellulose of concern. A convenient .cif was available as supplemental material in French (2014); .cifs are routinely reported for small molecules and can be created if the space group and atomic coordinates are available.

Figure 2.10 is a synchrotron X-ray pattern from a more typical cotton fiber sample, colorized in the XRD2DScan software. Because individual cotton fibers are so small, samples are composed of hundreds of fibers, bundled and combed to be parallel. In this image, the fibers are vertical, and there is a black line at 7:00 o'clock that is a shadow from the "beam stop" support. (The beam stop is typically made of lead and prevents the powerful main beam from destroying the detector.) Scans of the equator and meridian are shown with the Miller indices for some of the peaks. The two scans are not to scale; the meridional scan is exaggerated so as to make the weak peaks more visible.

Patterns similar to Fig. 2.10 but from conventional laboratory X-ray diffraction systems have been available for many years (Clark 1930; Howsman and Sisson 1954). The Clark article explains that cotton was typically not the subject of X-ray work for fundamental understanding because of the less perfect orientation of the microfibrils (micelles) to the fiber axis. Many of the basic principles were described in the Howsman and Sisson chapter, and a following chapter in the same book (Hock 1954) specifically addresses the varying orientations for the secondary wall microfibrils. The long arcs were considered to result from a distribution of microfibril orientations similar to the S1 and S2 layers in the individual cells in wood. There are some advantages in the modern toolkit for improved understanding, especially the availability of synchrotron X-ray radiation that was powerful enough to investigate single fibers as previously shown in Fig. 2.8. A more thorough study of the cell wall structure is needed, but the tools are now available to solidify the

[2]CAMD, the Center for Advanced Microstructures and Devices, is a facility of Louisiana State University, Baton Rouge, LA.

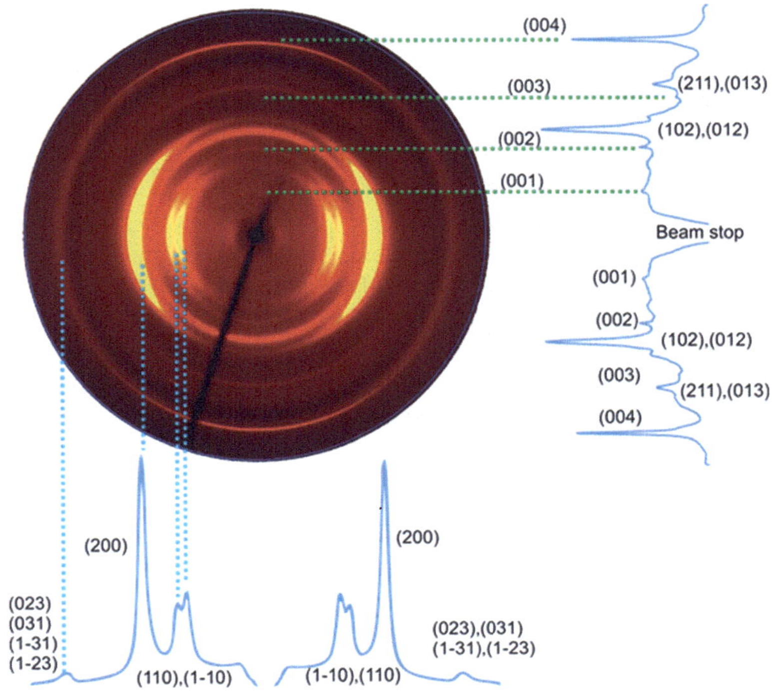

Fig. 2.10 Cotton fiber bundle diffraction pattern from the CAMD synchrotron. The pattern was colorized by the XRD2DScan software. Superimposed are scans of the equator (bottom) and meridian (right side) with the corresponding peak labels (Miller indices). The meridian scan, in particular, gives the impression that there are numerous peaks just above the background. The meridional scan gave a sharp peak for the (004) reflection, but at the same 2θ value, the equatorial scan peak was weaker and diffuse, partly due to the presence of four different peaks of weak but significant intensity. Some even weaker peaks were not included in the labels. The green dashed lines point to the (00 l) reflections, all of which are visible in the original pattern as well as on the scan

indications from these patterns that a single microfibril orientation dominates at a given position along the length of the fiber.

Additionally, Fig. 2.10, also with synchrotron radiation, has better resolution. The equatorial scans of the pattern (with ImageJ) show broader peaks than the well-resolved (004) peak on the meridional scan, consistent with the electron diffraction and simulated patterns in Fig. 2.7. The (002) reflection is also sharp but quite weak. The published reproductions will probably not show the (001) and (003) lines that are barely visible on the original images and just bumps on the tracings. Their appearance on the experimental pattern is not expected because those reflections are disallowed by the $P2_1$ space group. One explanation is that the forbidden reflections could arise from the finite length of the model crystallites. Figure 2.7b shows some similar intensity for the (003) reflection despite the crystal model conforming to the

$P2_1$ space group. Support for arguments that the cellulose chains in cotton conform to twofold screw-axis symmetry arises from the lack of improvement in the fit between observed and calculated diffraction intensities for tunicate cellulose (Nishyama et al. 2002). NMR studies show that there are two unique glucose residues in tunicate cellulose, but they could either be in the same chain in which case they would not have screw-axis symmetry or in two separate chains as proposed by the X-ray study (Kono et al. 2002) (It now seems that there are numerous unique chains in cellulose, based on NMR).

The XRD2DScan software offers other facilities of interest to cotton research. Figure 2.11 shows a scan after conversion of the arcs in Fig. 2.10 to full diffraction rings. This circularization of the arcs allows the data to be treated in the same manner as a conventional X-ray powder diffraction pattern. The simulated powder pattern for a perfect, randomly oriented tunicate cellulose powder composed of large crystals is also shown below the observed pattern, as are all of the potential peaks indicated by the magenta lines. This constitutes a transmission powder pattern, as compared to a reflection mode pattern. The two types of patterns have inherently opposite preferred orientation preferences. Reflection mode patterns typically have weak meridional and near-meridional reflections because the fiber fragments in a powdered sample tend to lie down in the plane of the sample holder. Therefore the

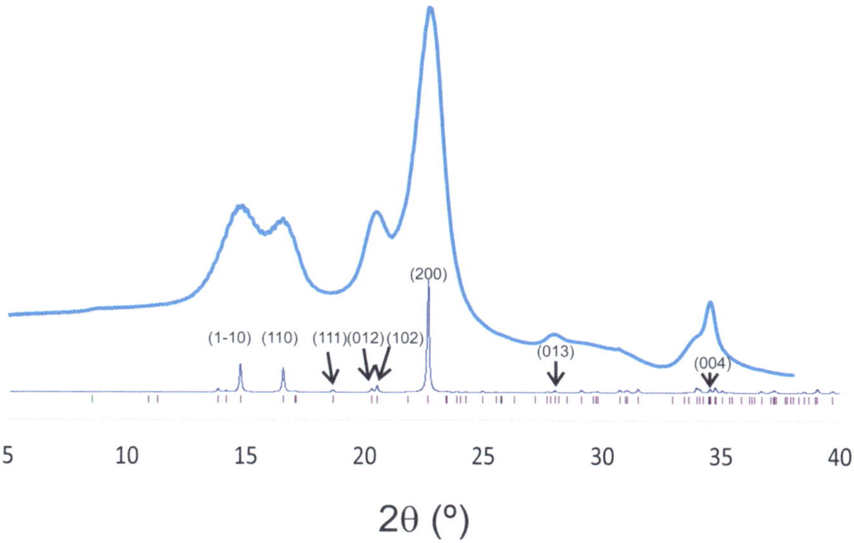

Fig. 2.11 This image contains the "circularized" diffraction intensities from Fig. 2.10, as well as the calculated intensities for cellulose I_β as given by the Mercury software, based on the Nishyama et al. (2002) structure. The circularized intensities from the fiber bundle are noticeably different from the intensities in Fig. 2.6a, in part because of the different sample orientation and transmission (Fig. 2.10) instead of reflection diffraction geometry (Fig. 2.6a). The positions of all predicted peaks are indicated by vertical magenta lines, along with the very sharp peaks (pwhm 0.1° 2θ) that would come from a good crystal greater than 100 nm in size

(002) and (004) planes are perpendicular to the sample holder plane, and it is difficult to get many fiber fragments into the Bragg condition for the meridional reflections. On the other hand, in transmission mode (Fig. 2.11), the oriented fragments can achieve the Bragg condition, and those intensities are enhanced. The observed intense reflections for the (102) and (012) reflections at 20° 2θ and the sharp (004) spike in the midst of the more diffuse peak at 34.6° 2θ are much stronger than indicated by the calculated pattern, because the sample is composed of oriented fibers.

The ability to separate the equatorial and meridional scans permits a better determination of the crystallite size. In the case of the meridional scan, the pwhm (0.4°), determined by the fityk program (Wojdyr 2010; http://fityk.nieto.pl/), converts to a crystallite length of about 212 Å. The equatorial scan had some background issues, but the pwhm sizes for the (1–10), (110), and (200) peaks were 1.84, 1.77, and 1.66°, corresponding, respectively, to crystallite sizes perpendicular to those planes of 49, 48, and 50 Å. Model crystals can then be based on these sizes. They constitute minimum values.

Besides the typical 2θ scans shown in Fig. 2.11, the XRD2DScan program can scan around a circle at a given 2θ value on a pattern such as Fig. 2.10. Figure 2.12 shows such an "azimuthal scan" or "psi scan." This is a way to measure the shapes of the arcs and thus the MFA. Most such scans for cotton involve the (200) reflection at about 22.7° 2θ, but this particular scan was for the (1–10) reflection (a band from 14.2 to 15.4° 2θ. There is less possibility from interference with other reflections on upper layer lines for the (1–10) reflection, but measurements on the pattern of

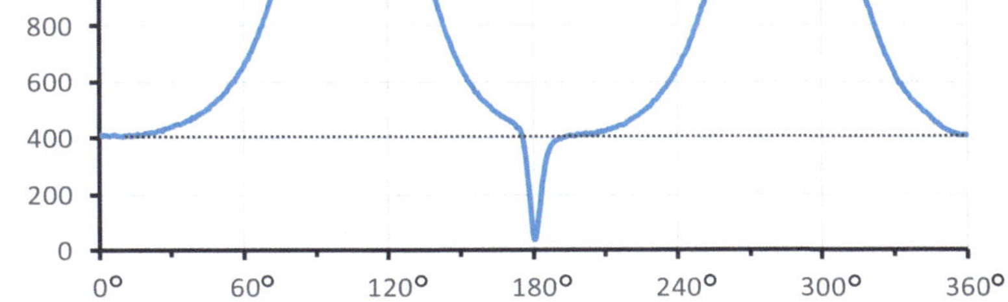

Fig. 2.12 "Azimuthal" or "psi" scan of the (1–10) diffraction arcs in Fig. 2.10. Although other measures of the MFA exist, the pwhm (here, 60°) is simple, and the value represents the extent of deviation from perfect alignment of the microfibrils. The dip at 180° is for the shadow of the beam stop support. From XRD2DScan

Fig. 2.10 for both the (1–10) and (200) reflections gave similar results, with peak widths at half height of about 60°.

2.4 Rietveld Method of Crystallinity Analysis

2.4.1 Problems with Current Methods

Park et al. (2010) summarized three common X-ray methods for determining cellulose crystallinity from powder diffraction patterns: peak height (Segal 1959), amorphous subtraction, and deconvolution. While all provide somewhat similar results, they all have ultimately disqualifying flaws. The Segal and amorphous subtraction methods attribute all intensity at approximately 18.5° 2θ to amorphous material. However, much of that observed intensity can often be attributed to crystalline material. There is not only a small but significant peak at that point (111) (see Fig. 2.11), but also there is substantial overlap between the tails of the much stronger peaks on either side that can be convincingly simulated based on the pseudo-Voigt peak shape. As progressively smaller but realistic crystallites are considered, the overlap becomes more and more severe (French and Santiago Cintrón 2013). The problem with the deconvolution method lies in the failure to account for all of the diffraction peaks that are present and the failure to link the calculated peak intensities and widths to known diffraction physics. The peak deconvolution method would be correct in principle if these faults were avoided. The approach that we recommend is called the Rietveld method (Young 1995), and it specifically avoids these problems. Of course, there are still problems, especially including the separation of "background" scattering from "amorphous" scattering. It is hoped that these problems, exacerbated by the limited amount of data from most cellulose samples, will be overcome in the coming years. If it turns out that not all of the desired information can be reliably extracted from practical samples, at least the limits on our ability to analyze the data could be established. At present, there are no established protocols for application of the Rietveld method to cellulose samples.

The original purpose of Rietveld analysis was to provide a method to solve crystal structures with neutron diffraction. Many molecules have not been crystallized in crystals large enough to analyze with neutron diffraction using conventional single-crystal methods, but they can still make suitable samples in the powder form. In a conventional single-crystal study, the idea is to match individual (*hkl*) reflection intensities calculated from a model with those of the corresponding observed spots. In a Rietveld approach, the idea is to match observed and calculated intensities at each increment of 2θ that is observed. This eliminates the need to separate overlapped reflections. To do that, it is necessary to consider the crystallite size, orientation, and multiple phases (allomorphs, other materials, and amorphous cellulose).

For cellulose, there is not nearly enough information in a powder diffraction pattern to solve for the *x*, *y*, and *z* coordinates of the atoms, but the secondary information, regarding, for example, crystallite size and amorphous content, are within the capability of the Rietveld method if the atomic coordinates are known. Knowledge of the coordinates, such as from the study by Nishyama et al. (2002), are read into the Rietveld software in the form of .cif files, along with the experimental intensity vs. 2θ. The ideal diffraction pattern from a cellulose crystal is easily calculated, and the question for the analysis is essentially "what modifications to the ideal structure are needed to match the observed experimental pattern?"

2.4.2 Sample Analysis of TM-1 Upland Cotton

Some instructive examples from a Rietveld analysis of a standard cotton sample are shown in Fig. 2.13. Such refinements are typically done with incremental increases in the number of fitting variables (Kim et al. 2017). The first example (a) shows the experimental and ideal patterns after a guess of the scale factor. Obviously missing is some accounting for the background. Additionally, the ideal peaks are much too sharp because they are for a crystal much larger (e.g., 1000 Å) than a cotton crystal. The second image (b) shows how the calculated peaks expand when a crystallite size of 60 Å is modeled by the calculation. The third example (c) shows the results after the first refinement that included a scale factor, a linear background, and the unit cell *a*-axis. The fourth example (d) shows the results from the Maud program

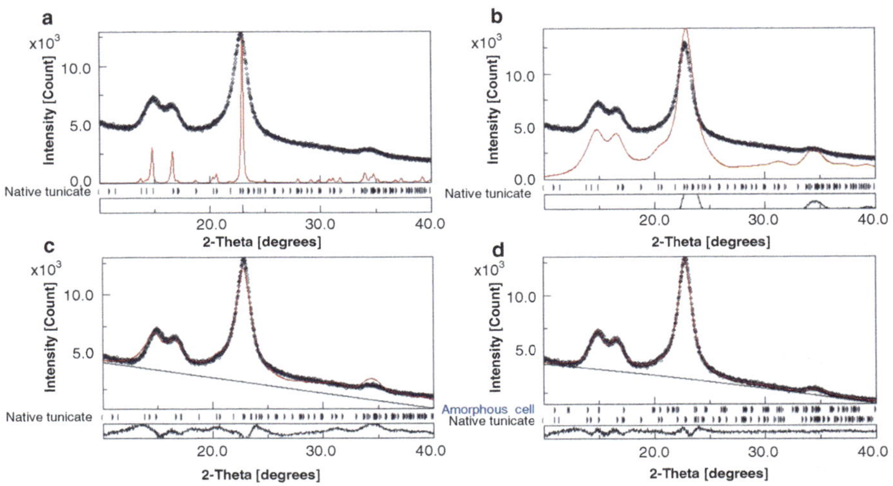

Fig. 2.13 (**a**) Observed pattern for TM-1, along with calculated intensities for an ideally large crystal (in red). (**b**) Calculated intensities changed to model a perfect crystal with 80 Å cross section. (**c**) Calculated intensities after first refinement, with calculated linear background, variable *a*-axis length, and scale factor. (**d**) After 13 more variables included in the refinement. From the Maud program

(Lutterotti, http://radiographema.eu) after inclusions of 13 more variables, added incrementally as indicated in Table 2.1. Some Rietveld refinements freeze the optimized variables before adding new variables, but in this project, all variables were refined once initially included.

Big improvements (reductions) in the residuals (the discrepancy indices *Rwp* and *Rb*) were obtained when the preferred orientation was modeled with March-Dollase factors for fiber orientation. The randomly oriented samples would have some intensity around 20° 2θ [the (102) and (012) reflections] and increased intensity near 34.6° 2θ. The difference between the randomly oriented ideal pattern and the experimental sample is particularly obvious in Fig. 2.13c at about 34.6°, before including the compensation for preferred orientation. Although the peak widths in the scans of Figs. 2.7a and 2.10 indicate a very non-isotropic crystallite shape (longer along the *c*-axis than the *a*- and *b*-axes), inclusion of six variable shape parameters (the isotropic size is not considered when the anisotropic shape parameters are used) did not produce such a dramatic drop in the *R*-values.

Another point regards the introduction of a second phase in refinement #5. The coordinates for cellulose II were read in, and its crystallite size was fixed at 12 Å. The calculated pattern from cellulose II for such a small crystallite size was shown to closely resemble that of exhaustively ball-milled cellulose (Nam et al. 2016), so this approach allows a refinement of a distinct amorphous fraction. The Maud program refined this to indicate a 77%/23% crystalline/amorphous composition, but Fig. 2.14b indicates that there is relatively little (about 9%) area under the curve for the amorphous material. We do not know the reason for this discrepancy, but we accept the smaller value based on photon counts from the spreadsheet rather than the Maud program dialogue. When the crystallite size of the second phase was allowed to refine, the crystallite size increased to 45 Å, suggesting the presence of some 5% of cellulose II and little, if any, amorphous material.

2.5 Summary

This chapter covered the crystal and molecular structure as well as some aspects of the supramolecular structure. Reference is made to some work initiated more than 80 years ago that is still not quite able to yield a sufficient understanding. The summary of various dimensions for the structural hierarchy could have been written many years ago, but our modern listing could be helpful. One finding herein that is potentially transformative for the cotton literature is that the layers of secondary wall at any given point along the fiber length may not be progressively changing. Another transformative point may be that the amount of scattered intensity that comes from a component of the fiber that resembles ball-milled cotton is very small.

Table 2.1 Rietveld refinement of TM-1 cotton

Refinement #	1	2	3	4	5	6	7
Added variable parameters	Linear background, scale, unit cell *a*-axis (Fig. 2.13c)	Quadratic background, unit cell γ, crystallite size	March-Dollase preferred orientation	*b*-axis	Including model amorphous as second phase	Non-isotropic shape	Refine size of second phase crystallites (Fig. 2.13d)
Total number of varied parameters	4	7	8	9	11	16	17
Sigma[a]	3.77	3.64	1.94	1.85	1.81	1.69	1.65
Rwp[a]	5.68	5.48	2.91	2.78	2.73	2.53	2.48
Rb or *Rp*[a]	4.38	4.15	2.34	2.27	2.22	2.03	1.97

[a]Sigma (or S) represents an indicator similar to the goodness of fit in single-crystal work, an indicator of the significance of the improved fit of the observed and calculated diffraction intensities as the number of variables is increased. *R*wp is the square root of the sum of the weighted discrepancies squared divided by the sum of the squares of the weighted observed intensities. *Rb* (or *Rp*) corresponds to the simple crystallographic *R* value, the sum of the absolute values of the discrepancies divided by the sum of the observed intensities. Unlike single-crystal work, these intensities are the values at each increment of 2θ instead of at the *hkl* reciprocal lattice points (Young 1995).

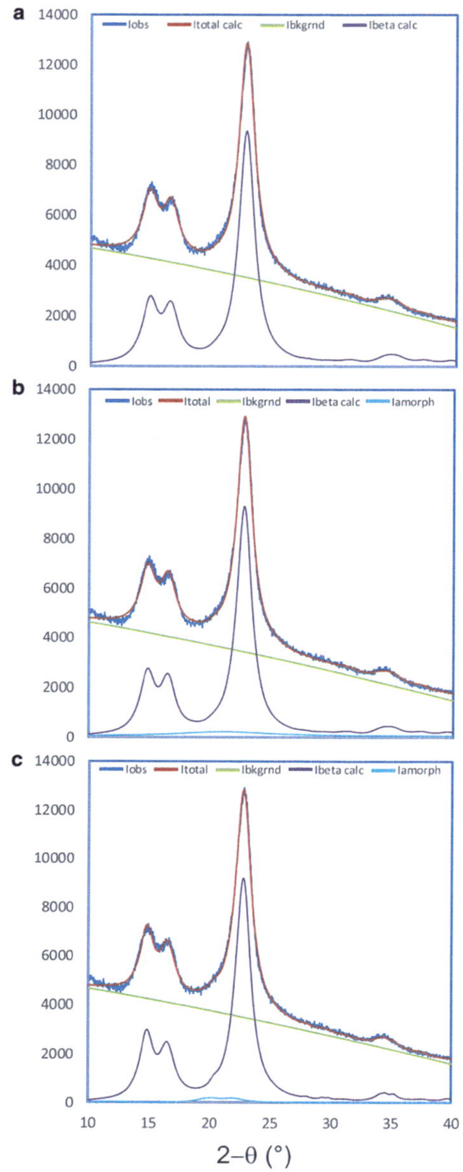

Fig. 2.14 Plots from the Maud program showing the fit of calculated components to the observed pattern. The key information is in the light blue line at the bottom of (**b**) and (**c**). The fit in (**a**) is visually nearly as good without a second phase such as amorphous (**b**) or crystalline II (**c**). The keys to the lines are Iobs, experimental intensity; Itotal, total calculated intensity; Ibeta calc, calculated intensity for crystalline native cellulose; Ibkgrnd, calculated background based on an optimized quadratic (second-order polynomial) function

Cutting-edge NMR studies have not yet been extensively applied to cotton, but it is clear that numerous problems will be attacked with those methods in the future.

Relatively new efforts that should have fruitful results include further work with single-fiber diffraction that can more clearly connect the convolutions with the changing microfiber orientation to help understand the secondary cell wall development. The extra resolution provided by synchrotron diffraction on bundles of cotton fiber can be effective for determining the effective crystallite dimensions. Also, the experimentally observed variable fiber elongation for different samples during strength testing could depend on the extent of misalignment of the microfibrils with the fiber axis. Finally, the ability to simulate the powder diffraction pattern for cotton samples can help to understand the experimentally determined powder patterns. The question becomes "what must be done to an ideal sample crystal to make its diffraction pattern resemble the experimental pattern?"

In this writer's experience, the presented Rietveld refinement of TM-1 cotton gave exceptional agreement between the observed and calculated patterns. The final variables refined were those related to the shape anisotropy clearly observed in the experimental fiber patterns (Figs. 2.7a and 2.10). That refinement fine-tuned the agreement between the observed and calculated data, but the values of the added variables were not totally convincing. Issues of the background and other factors are not completely resolved. Yet, it is clear that the current popular methods (peak height, amorphous subtraction, and deconvolution) have substantial flaws, starting with the faulty assignment of amorphous contributions based on what appears to be overlap of the tails of the diffraction peaks. That overlap is very dependent on the crystallite size. Size effects can also be a factor in various experimental studies of crystallinity, such as deuteration or spectroscopy.

References

Abidi N, Manike M (2017) X-ray diffraction and FTIR investigations of cellulose deposition during cotton fiber development. Text Res J 88(7):719. https://doi.org/10.1177/0040517516688634

Abidi N, Cabrales L, Haigler CH (2014) Changes in the cell wall and cellulose content of developing cotton fibers investigated by FTIR spectroscopy. Carbohydr Polym 100:9–16

Bader RFW (1990) Atoms in molecules: a quantum theory, International series of monographs on chemistry. Oxford Clarendon Press, New York

Clark GL (1930) Cellulose as it is completely revealed by X-rays. Special application to the growth and classification of cotton, the structure of wood, and the manufacture of rayon. Ind Eng Chem 22:474–487

Dri FL, Hector LG Jr, Moon RJ, Zavattieri PD (2013) Anisotropy of the elastic properties of crystalline cellulose Iβ from first principles density functional theory with Van der Waals interactions. Cellulose 20:2703–2718

Fernandes AN, Thomas LH, Altaner CM, Callow P, Forsyth VT, Apperley DC, Kennedy CJ, Jarvis MC (2011) Nanostructure of cellulose microfibrils in spruce wood. Proc Natl Acad Sci U S A 108:E1195–E1203

French AD (2014) Idealized powder diffraction patterns for cellulose polymorphs. Cellulose 21:885–896

French AD, Santiago Cintrón M (2013) Cellulose polymorphy, crystallite size, and the segal crystallinity index. Cellulose 20:583–588

French AD, Concha M, Dowd MK, Stevens ED (2014) Electron (charge) density studies of cellulose models. Cellulose 21:1051–1063

Gessler K, Krauss N, Steiner T, Betzel B, Sarko A, Saenger W (1995) β-D-Cellotetraose hemihydrate as a structural model for cellulose II. An X-ray diffraction study. J Am Chem Soc 117:11397–11406

Goynes WR Jr. (2005) In search of the mystic cotton fiber maturity: a view from the microscope. Proceeding of Beltwide Cotton Conference, New Orleans, Louisiana, pp 2212–2218

Haigler CH, Rao NR, Roberts EM, Huang J-Y, Upchurch DR, Trolinder NL (1991) Cultured ovules as models for cotton fiber development under low temperatures. Plant Physiol 95:88–96
Han YJ, Cho Y-J, Lambert WE, Bragg CK (1998) Identification and measurement of convolutions in cotton fiber using image analysis. Artif Intell Rev 12:201–211
Hock CW (1954) Microscopic structure. In: Ott E, Spurlin HM (eds) Cellulose and cellulose derivatives, part 1. Interscience, New York, pp 347–392
Howsman JA, Sisson WA (1954) Submicroscopic structure. In: Ott E, Spurlin HM (eds) Cellulose and cellulose derivatives, part 1. Interscience, New York, pp 231–346
Kalenius E, Kekalainen T, Neitola R, Beyeh K, Rissanen K, Vainiotalo P (2008) Size- and structure selective noncovalent recognition of saccharides by tetraethyl and tetraphenyl resorcinarenes in the gas phase. Chem Eur J 14:5220–5228
Kim HJ et al (2017) Comparative physical and chemical analyses of cotton fibers from two near isogenic upland lines differing in fiber wall thickness. Cellulose 24:2385–2401
Kim HJ, Liu Y, French AD, Lee CM, Kim SH (2018) Comparisons and validation of Fourier transform infrared spectroscopic methods for monitoring secondary cell wall cellulose from cotton fibers. Cellulose 25:49–64
Kono H, Unoki S, Shikano T, Fujiwara M, Erata T, Takai M (2002) CP/MAS ^{13}C NMR study of cellulose and cellulose derivatives. 1. Complete assignment of the CP/MAS ^{13}C NMR spectrum of the native cellulose. J Am Chem Soc 124:7506–7511
Krässig H, Käppner W (1961) The morphological units in cotton linters. Makromol Chem 44:1–7
Lee CM, Kubicki JD, Fan B, Zhong L, Jarvic MC, Kim SH (2015) Hydrogen-bonding network and OH stretch vibration of cellulose: comparison of computational modeling with polarized IR and SFG spectra. J Phys Chem B 119:15138–15149
Liu Y, Thibodeaux D, Gamble G, Bauer P, Van Derveer D (2012) Comparative investigation of Fourier transform infrared (FT-IR) spectroscopy and X-ray diffraction (XRD) in the determination of cotton fiber crystallinity. Appl Spectrosc 66:983–986
Liu J, Kim JI, Cusumano JC, Chapple C, Venugopalan N, Fischetti RF, Makowski L (2016) The impact of alterations in lignin deposition on cellulose organization of the plant cell wall. Biotechnol Biofuels 9:126–142
Nam S, French AD, Condon BD, Concha M (2016) Segal crystallinity index revisited by the simulation of X-ray diffraction patterns of cotton cellulose Iβ and cellulose II. Carbohydr Polym 135:1–9
Nishiyama Y, Kuga S, Wada M, Takeshi O (1997) Cellulose microcrystal film of high uniaxial orientation. Macromolecules 30:6395–6397
Nishiyama Y, Johnson GP, French AD (2012) Diffraction from nonperiodic models of cellulose crystals. Cellulose 19:319–336
Nishyama Y, Langan P, Chanzy H (2002) Crystal structure and hydrogen-bonding system in cellulose Iβ from synchrotron X-ray and neutron fiber diffraction. J Am Chem Soc 124:9074–9082
Park S, Baker JO, Himmel ME, Parrilla PA, Johnson DK (2010) Cellulose crystallinity index: measurement techniques and their impact on interpreting cellulase performance. Biotechnol Biofuels 3:10
Purves CB (1943) A historical survey. In: Ott E (ed) Cellulose and cellulose derivatives. Interscience, New York, pp 29–53
Rajasekaran K, Muir AJ, Inbger BF, French AD (2006) A dehydration method for immature or wet cotton fibers for light and electron microscopy. Text Res J 76:514–518
Santiago Cintrón M, Johnson GP, French AD (2011) Young's modulus calculations for cellulose Iβ by MM3 and quantum mechanics. Cellulose 18:505–516
Segal L, Creely JJ, Martin AE, Conrad CM (1959) An empirical method for estimating the degree of crystallinity of native cellulose using the X-Ray diffractometer. Text Res J 29(10):786–794
Sisson WA (1935) X-ray studies of crystallite orientation in cellulose fibers. Ind Eng Chem 27:51–56
Sponsler OL, Dore WH (1926) The structure of ramie cellulose as derived from X-ray data, Colloid symposium monograph, vol IV. Chemical Catalog, New York, pp 174–204

Stevens ED, Dowd MK, Johnson GP, French AD (2010) Experimental and theoretical electron density distribution of α,α-trehalose dihydrate. Carbohydr Res 345:1469–1481

Sugiyama J, Vuong R, Chanzy H (1991) Electron diffraction study on the two crystalline phases occurring in native cellulose from an algal cell wall. Macromolecules 24:4168–4175

Wakelyn PJ, Bertoniere NR, French AD, Thibodeaux DP, Triplett BA, Rousselle M-A, Goynes WR Jr, Edwards JV, Hunter L, McAlister DD, Gamble GR (2007) Cotton fiber chemistry and technology. CRC Press, Boca Raton

Whitaker PM, Nieduszynski IA, Atkins EDT (1974) Structural aspects of soda-cellulose II. Polymer 15:125–127

Wojdyr M (2010) Fityk: a general-purpose peak fitting program. J Appl Crystallogr 43:1126–1128

Yang H, Wang T, Oehme D, Petridis L, Hong M, Kubicki JD (2018) Structural factors affecting ^{13}C NMR chemical shifts of cellulose: a computational study. Cellulose 25:23–36

Young RA (1995) Introduction to the Rietveld method. In: Young RA (ed) The Rietveld method. Oxford University Press, Oxford

Zugenmaier P (2008) Crystalline cellulose and cellulose derivatives. Characterization and structures. Springer, Berlin

Chapter 3
Physical Properties of Cotton Fiber and Their Measurement

Christopher D. Delhom, Brendan Kelly, and Vikki Martin

3.1 Introduction

There are numerous physical properties that influence the quality of cotton. These properties, e.g., length, strength, color, micronaire, fineness, and maturity, must all be quantified for different objectives in research and industrial applications. Indeed, cotton quality means many different things depending on where in the supply chain one is located. For example, the various fiber quality parameters have differing levels of importance depending on the products being manufactured and their subsequent end use. One critical juncture for evaluating cotton fiber properties is cotton classification. Cotton classification is the process of assigning values to allow the market to properly value ginned lint, facilitating the trade of cotton. Classification focuses on the traditional cotton properties of color, length, strength, micronaire, and non-lint content (i.e., trash). There are many additional parameters that can be measured and utilized to understand a cotton sample; these include the distributions of the bulk averaged properties measured in classification, fiber maturity and fineness, single-fiber tensile properties, neps, and short fiber content.

While these parameters are not used in the current classification system, they are frequently utilized in research. For example, breeders use fiber quality parameters

---

C. D. Delhom (✉)
Cotton Structure and Quality Research Unit, USDA-ARS, Southern Regional Research Center, New Orleans, LA, USA
e-mail: chris.delhom@ars.usda.gov

B. Kelly
Texas Tech University, Lubbock, TX, USA

Texas A&M Agrilife Research, Lubbock, TX, USA

V. Martin
Cotton Incorporated, Cary, NC, USA

D. D. Fang (ed.), *Cotton Fiber: Physics, Chemistry and Biology*,
https://doi.org/10.1007/978-3-030-00871-0_3

to estimate the potential spinning performance, and they often need to separate fiber qualities into their heritable components.

Originally, cotton classification was a manual process in which a trained classer visually assessed lint color and non-lint content, so-called leaf grade. The classer would manually draw a fiber sample from the classing sample and manipulate the subsample into a parallel beard of fibers, and then, the classer could visually estimate the staple length of the fiber in 32nd of an inch while visually judging the bulk sample for overall color and non-lint content. In the last quarter of the twentieth century, technology advances were made to quantify cotton quality for classification and research purposes. Cotton classification is now performed by Standardized Instruments for Testing of Cotton (SITC), the most common of which is the Uster High Volume Instrument (HVI) shown in Fig. 3.1. While there are many automated techniques available for measuring the physical properties of cotton, the manual reference methods are still relevant as well.

A key factor in all physical property characterization of cotton is that the environment can impact the results and the testing should thus be carried out in standard atmospheric conditions of 21 ± 1 °C and 65 ± 2% relative humidity as specified in ASTM D1776 (2016). Cotton must be allowed to equilibrate to the environment prior to testing; however, the final moisture content and fiber properties are also potentially impacted by the conditions the cotton was exposed to prior to equilibration at standard conditions due to hysteresis (Urquhart and Eckersall 2008).

The various demands from the industry have led to the development of many test methods and instruments to measure the physical properties of cotton (Table 3.1). This chapter provides an overview of the most important fiber physical characteristics and how they are measured.

Fig. 3.1 Uster HVI 1000

Table 3.1 Key physical properties of cotton fiber

Property	Included in instrument-based classing? (Y or N)	Units	ASTM standard(s)
Length	Y	mm (in)	D1440, D5867
Strength	Y	g/tex	D1445, D5867
Micronaire	Y	Unitless	D1448, D5867
Fineness	N	mtex	D1577
Maturity	N	Unitless	D1464
Color	Y	Rd, +b	
Neps	N	Count/gram	D5866
Fiber cohesion	N	μN/tex	D2612, D4120

Fig. 3.2 A staple pull by a trained hand classer

3.2 Length

Length is the best overall indicator of fiber quality and plays a key role in determining yarn quality and processing performance. As cotton is a natural product, the length of fibers in a bale is quite variable. Traditionally, a single number is reported for length which is referred to as the staple length. In the early years, staple length was determined by a cotton classer. The classer aligned the fibers and ensured only continuous fibers remained in the handheld subsample, which was then judged to the nearest 32nd of an inch (0.79 mm) visually (Fig. 3.2).

In modern classing instruments, the most common single length parameter is the upper half mean length (UHML), which is defined to be the mean length by number of the longest 50% of the fibers in a sample by weight. This measurement is considered the equivalent of traditional staple length. Length parameters can be calculated on a number basis (calculations are performed based on the number of fibers measured) or a weight basis (calculations are performed based on the weight of the

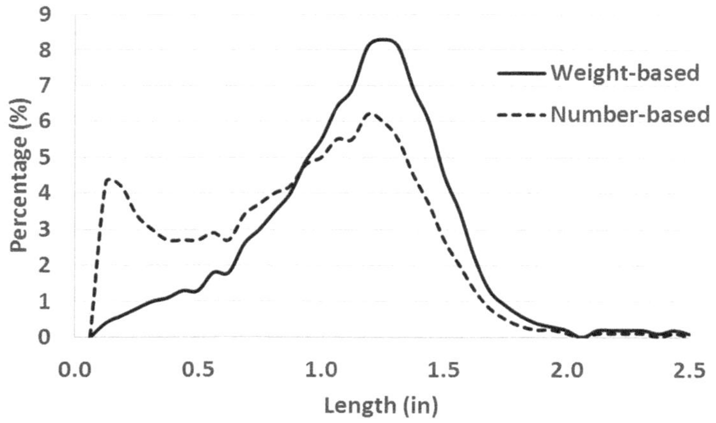

Fig. 3.3 AFIS-generated weight and number-based length distributions for a single cotton

sample measured). Shorter fibers generally weigh less than longer fibers, and therefore the length distributions have different shapes when based on weight versus number (Fig. 3.3).

3.2.1 *Array Method*

The reference method for length measurement is an array method. In the Western world, the array method most commonly used is the Suter-Webb Array (ASTM D1440 2012) and similar methods such as the German DIN 53806 (DIN 53806:1970-02 1970). For these array methods, an operator manually parallelizes the fibers in an approximately 80 mg sample and removes all non-lint content. The operator places this sample on one needle bed of a dual-bank comb sorter (Fig. 3.4). Each bed of needles in the Suter-Webb sorter is composed of a series of combs that are able to drop in a sequential order from the front to back. A pair of wide forceps is used to grab fibers which are protruding from the first comb in the needle bed, and this sample of fibers is moved to the needle bed on the opposite bank of the sorter. The first comb in the bed is then lowered, and the process is repeated until the sample has been transferred from the original bank to the opposite bank. Dropping each comb and only pinching and transferring fibers protruding from the now dropped comb. The instrument is pivoted, and a similar process is used to drop combs and transfer the sample from one needle bed to the other. This process effectively sorts the fibers into length groups, determined by the dropping of each comb. The fibers from each length group are weighed separately allowing a weight-based distribution and associated length parameters to be calculated.

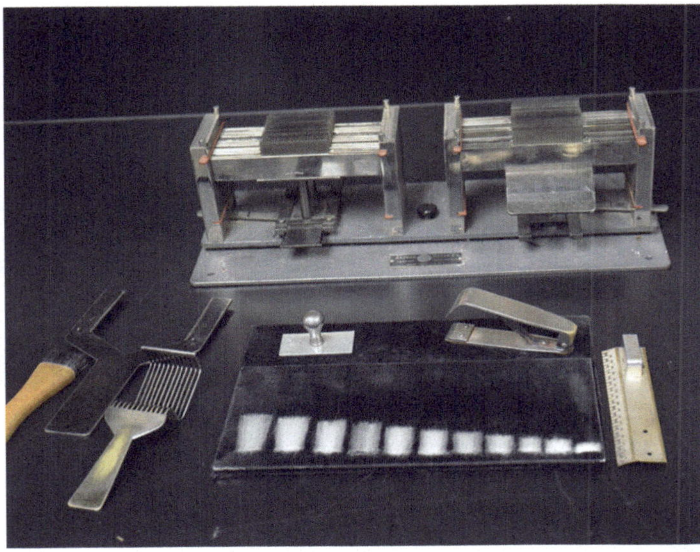

Fig. 3.4 Suter-Webb Array and equipment to prepare the array (dual-bank comb sorter, wide forceps, velvet board, and ruler)

In practice, the array method, and the closely related Chinese roller length analyzer method (GB/T6098.1-1985 1985), requires multiple skilled operators to cross-check each sample. The array methods are highly accurate when conducted by experienced operators. However, these methods are subject to operator bias and are time-consuming needing approximately 2 h per test (Bargeron 1986).

3.2.2 *Almeter*

The Peyer Texlab Almeter was introduced originally to measure the length distribution of wool. However, in the 1980s a modified version was made available for the measurement of cotton length distribution (Bargeron 1986). The Peyer Texlab utilized the Fibroliner (Fig. 3.5) to mechanically align the fibers into an end-aligned fiber beard. This end-aligned bundle was transferred to the Almeter (Fig. 3.6) for scanning with a capacitive sensor system (Fig. 3.7). Although the fibers may still contain crimp (kinks along the length of the fiber due to fiber development processes), the full length of the fibers is measured allowing this method to produce a full-length distribution. The Almeter method is considerably faster than the array methods. This method is not without disadvantages. Due to within-sample variation in fiber linear density, it is unclear whether the Almeter distribution, based on variation in capacitance, is characterizing variation in length by number or by weight.

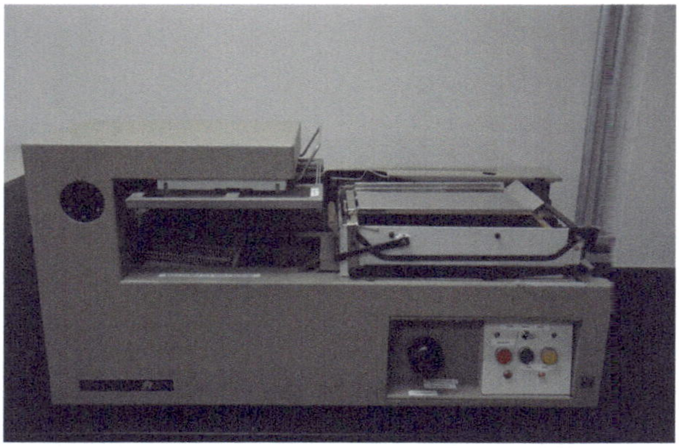

Fig. 3.5 Peyer Fibroliner FL-101

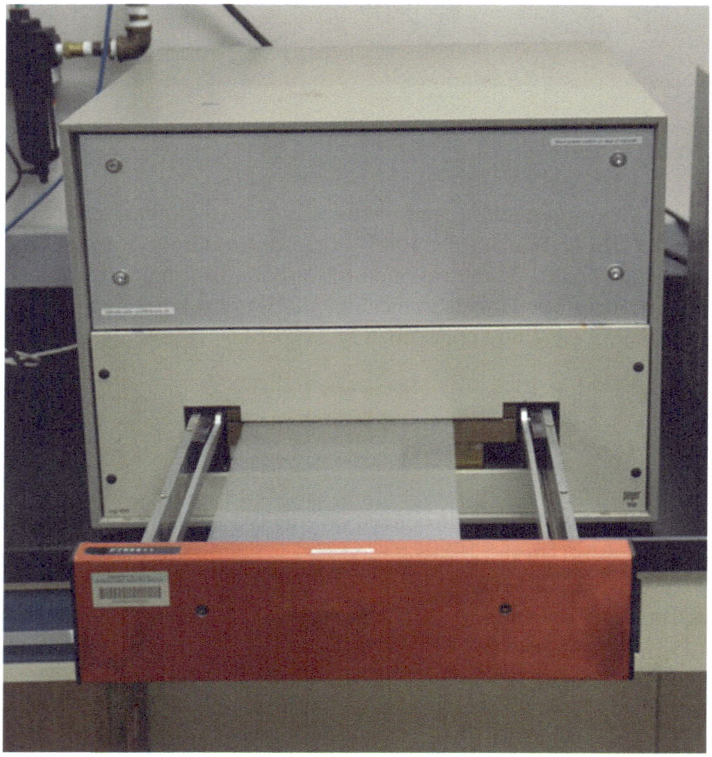

Fig. 3.6 Peyer Almeter AL-101 as updated by texma.org AG (Oberglatt, Switzerland)

Fig. 3.7 Almeter foil bed loaded with an end-aligned fiber beard produced with the Peyer Fibroliner. The sample is head between the foils and advanced into sensitive capacitor plates that are used to detect variation in the fiber beard

3.2.3 Standardized Instruments for Testing of Cotton (SITC)

The HVI (Uster Technologies, Knoxville, TN) and similar instruments, such as those produced by Premier Evolvics (Coimbatore, India), MAG Solvics (Coimbatore, India), Textechno (Moenchengladbach, Germany), and Loepfe Brothers (Wetzikon, Switzerland), use the basic principles of the fibrograph to measure cotton length. The fibrograph is an approach that measures light attenuation along a beard of fibers held in a clamp to produce a fibrogram (Hertel 1940). A fibrogram is a staple diagram created when the fibers are clamped on one end in a randomly distributed cotton sample, such as what occurs in processing. It is important to note that the clamp is applied randomly along the length of some fibers in the fibrograph method. An example of a fibrogram is shown in Fig. 3.8. A staple diagram is a plot of the cumulative length distribution when the fibers are edge-aligned, an arrangement that does not relate to processing nor occur outside of the laboratory. These systems do not characterize the complete within-sample distribution of fiber length, as the instrument is unable to measure the portion of the fibers held in the clamp (Fig. 3.9).

The above standardized instruments are highly automated, repeatable, and reasonably fast with testing times under 30 s per sample. Although based on the fibrogram principle, these instruments generally report limited length parameters with UHML, uniformity index (UI, the ratio of the UHML to the mean length), and short fiber index (SFI) being the most common parameters reported. It should be noted that UHML and UI are known values, while SFI is calculated by proprietary algorithms from the instrument manufacturers because the short fibers are located in the unmeasured portion of the fiber beard held in the comb. Short fiber index is not a direct measure of short fiber content and has limited use in research.

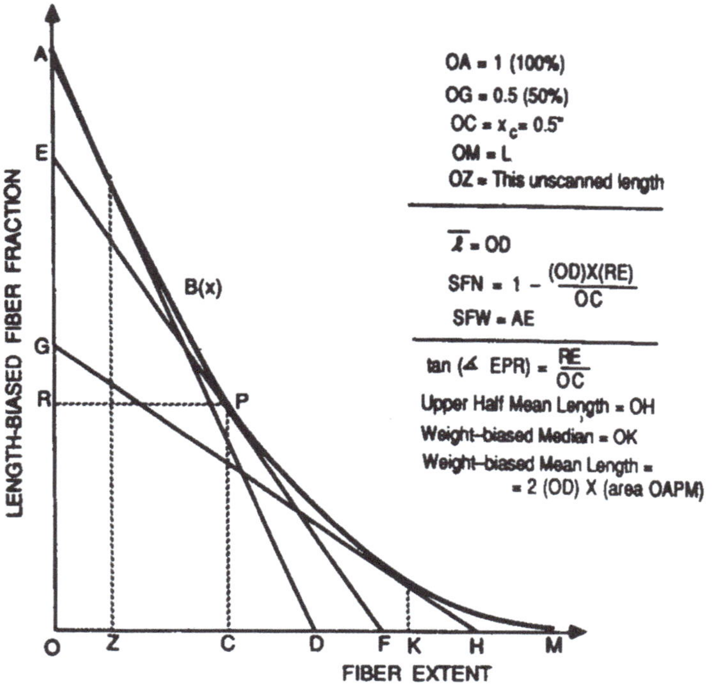

Fig. 3.8 Example of a fibrogram (Zeidman et al. 1991)

Fig. 3.9 Fiber beard being advanced into the HVI digital fibrogram device. The red light will be used to measure light attenuation of the fiber beard

3.2.4 Advanced Fiber Information System (AFIS)

The AFIS (Uster Technologies, Charlotte, NC) utilizes single-fiber measurements made by an electro-optical sensor (Bragg and Shofner 1993) (Fig. 3.10). The operator prepares a 0.5 g fiber sample into a 30 cm sliver which is fed into a pinned cylinder fiber individualizer. The fibers individualized from the bundle are pulled into an accelerated airstream and presented to electro-optical sensors which utilize time of flight to measure individual fiber lengths.

Not every fiber in the sample is measured by the AFIS sensors. Fiber hooks can result in the rejection of the individual fiber measurement. The airflow used to present the fiber to the AFIS sensors can partially remove some crimp, but the physical features of many fibers in the prepared sliver are not quantified by the AFIS system. Laboratory protocols require at least 3000–5000 fibers to be measured from each sliver and call for 3–5 slivers per samples. The results obtained from these 9000–15,000 fibers are used to generate a relative frequency distribution of fiber length (Fig. 3.3), maturity, and fineness. A weight-based distribution is provided by AFIS, but fibers are not weighed by the system. Instead, this distribution is calculated from the relative frequency distribution assuming a constant linear density within the sample. While the AFIS length by weight distribution may have some applications, the assumption of a constant linear density within a sample of cotton is not accurate. Operation of the AFIS instrument requires skilled sample preparation, as fibers may be broken during sliver formation, altering the within-sample distribution of fiber length. The fiber individualizer within the AFIS instrument is also aggressive and can break fibers, altering the within-sample distribution of fiber length.

The AFIS is relatively fast, measuring the physical properties of 3000 fibers in 2–3 min per sample, and it provides a considerable amount of data but is too slow

Fig. 3.10 Uster AFIS PRO

for large-scale uses such as cotton classification. Although a complete length histogram, by both weight and number, is available from the instrument, summary parameters of mean length (ML), short fiber content (SFC), upper quartile length (UQL), and the coefficient of variation of the length (CV) are used in most applications. The AFIS has become an important tool for breeders and spinners due to the detailed information produced and the relative speed and repeatability compared with other instruments (Cui et al. 1998; Thibodeaux et al. 2008b; Cai et al. 2013).

3.3 Strength

Fiber strength is an important parameter of fiber quality. Weak fibers will break during processing resulting in shorter fibers that result in lower economic value and industrial utility. The nature of cotton fiber strength is not completely understood. Various theories relate fiber strength to the development of primary cell wall, secondary cell wall, and convolutions (Hebert 1975). The strength of cotton fibers has been shown to vary with boll position, seed position, and fiber length (Liu et al. 2001). The strength of cotton fibers is measured both on individual fibers and bundles of fibers. These measurements do not serve the same purpose and one cannot replace the other. While bundle tensile properties are useful for predicting yarn tensile properties, individual fiber tensile properties are used to characterize within-sample variation in fiber strength and potential fiber breakage during processing.

Single-fiber tensile properties are more time-consuming to measure than bundle fiber strength but enable distributions for tensile properties to be determined within a sample. Tensile properties are naturally variable within a sample. Most individual fiber tensile property testers that are used to capture this natural variation require the operator to load individual fibers manually into the instrument. Proper characterization of the distribution of variation requires at least hundreds of fibers to be broken.

The raw breaking force can provide important information, particularly when working with individual fibers. However, in addition to the intrinsic material strength, breaking force depends on the amount of material present (bundle testing and individual fiber testing) or the number of fibers being broken (bundle testing). In order to account for the potential differences in tensile properties due to variation in the bundle mass, fiber strength from bundle tensile property tests are typically expressed as bundle tenacity. Bundle tenacity is the breaking force of the bundle normalized by the mass of the bundle. Individual fiber tenacity can be derived in a similar manner, normalizing individual fiber breaking force by the fiber linear density. Bundle tenacity is typically 50% lower than single-fiber tenacity, largely due to variation in fiber elongation (Sasser et al. 1991; Frydrych 1995).

Single-fiber strength measurement techniques include the Mantis (Uster Technologies, Knoxville, TN) and Favimat (Textechno, Moenchengladbach, Germany). Reference methods for bundle strength are the Pressley and Stelometer methods, while automated bundle strength testing is carried out by a variety of SITCs such as the Uster HVI and Textechno Fibrotest.

3.3.1 Single-Fiber Tensile Testing

Large volume testing of single cotton fibers was made easier with the development of the Mantis in the early 1990s. The Mantis allowed for higher-throughput testing by assisting the operator in mounting fibers and digitally recording the results. Mantis testing is carried out in a horizontal plane with the operator holding fibers by their midsection and affixing one end of the fiber in a clamp. Suction was then applied by the instrument to pull the other end of the fiber straight prior to engaging the second clamp. Breaking force was normalized using the ribbon width of the fiber, which was measured via light projection. A gauge length of 3.2 mm (1/8 in) is used in Mantis tensile property testing (Hebert et al. 1995). This instrument is no longer in production but is still in use at a few locations.

The Favimat instrument (Fig. 3.11) was created to test the tensile properties of many different fiber types, natural or man-made, and specifically used on cotton since the early 2000s (Foulk and McAlister 2002). The Favimat utilizes a vertical axis in which the fibers are mounted in the upper clamp and then the lower clamp is closed. A small weighted clip is attached to the lower end of the fiber to aid in holding the fiber clamp and removing some crimp (Fig. 3.15). The Favimat has an adjustable gauge length and pretension and extension rate all of which may influence the testing results. Tenacity is breaking force normalized by the linear density

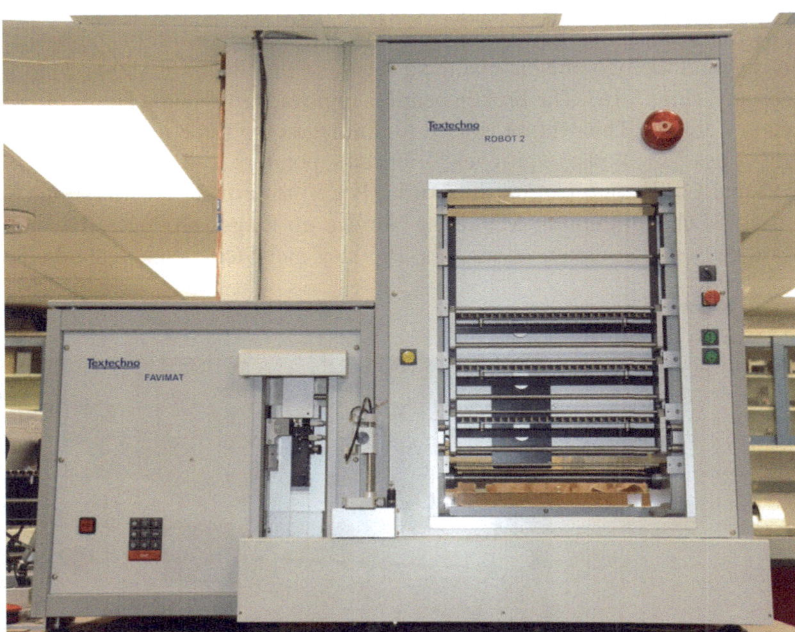

Fig. 3.11 Favimat single-fiber tester by Textechno H. Stein GmbH

of the fiber. Linear density is either manually entered or determined by the vibroscopic method (ASTM D1577 2016). If the vibroscopic method is being utilized, the minimum gauge length is 10 mm. This limits the length of fibers which can be tested. To increase productivity, the Favimat instrument can be loaded with the assistance of a robotic arm which transfers prepared fibers from a magazine into the measurement apparatus.

3.3.2 Flat Bundle Fiber Tensile Testing

The Pressley flat bundle tester (Fig. 3.12) was developed in 1939 and published as an ASTM standard in 1952 (Orr et al. 1955). The Pressley method utilized a flat bundle of cotton fibers placed in a clamp and broken using a sliding weight on an inclined plane. The Pressley method originally utilized a 0 mm (0 in) gauge length but was later equipped with an optional 3.2 mm (1/8 in) spacer plate. In the 1950s the Stelometer tester was introduced with an optional 0 mm and 3.2 mm gauge length (Figs. 3.13, 3.14, and 3.15). The Stelometer differs from the Pressley in the use of a pendulum and oil dampener to provide the breaking force.

The flat bundle testers utilize a fiber beard prepared in a similar manner to the array methods, in which the operator removes the non-lint content and combs the fibers into a beard. A special mechanism is utilized to eliminate crimp from the bundle, and a torque limiting clamp ensures consistent pressure is applied to the beard by the leather-lined jaws. The operator shaves the portion of the beard that extends beyond the two outside clamp surfaces, so that fibers of known length are being broken (Fig. 3.16). The broken beard is removed from both clamps after testing and is weighed. This weight is used to normalize the force to break and obtain tenacity. These methods require experienced operators and the use of USDA International Calibration Cotton Standards (ICCS) to provide a calibration. As of 2011 the USDA Agricultural Marketing Service no longer produces ICC cottons with Pressley or Stelometer values. The Pressley and Stelometer methods do not

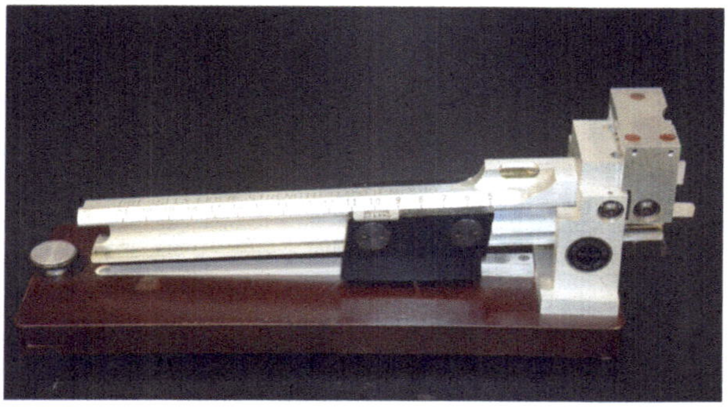

Fig. 3.12 Pressley flat bundle fiber strength tester

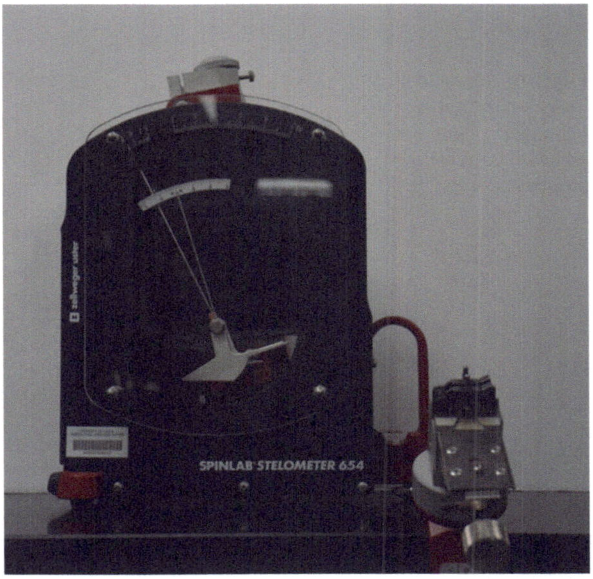

Fig. 3.13 Spinlab Stelometer 654

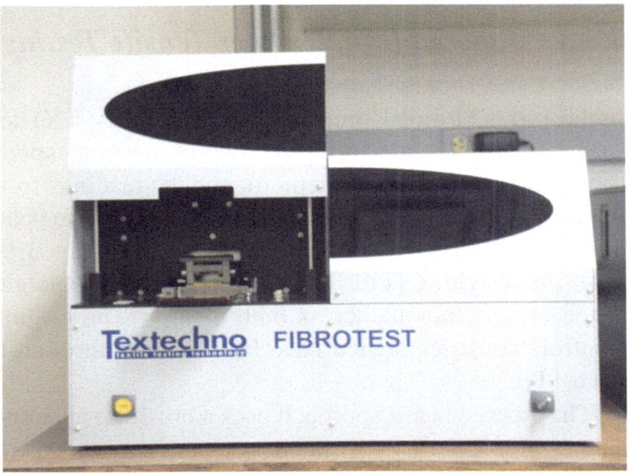

Fig. 3.14 Fibrotest by Textechno H. Stein GmbH

Fig. 3.15 Single cotton fiber being loaded into a Favimat single-fiber tester. A weighted fiber clamp is attached to the bottom of the fiber

Fig. 3.16 Flat bundle of fibers loaded into a Stelometer clamp being prepared for tensile property testing

provide the same results (Rouse 1964), and neither method provides identical results to SITCs (Sasser et al. 1991).

3.3.3 Tapered Bundle Fiber Tensile Testing

In the 1960s Motion Control Industries (Dallas, TX) developed an automated high-speed strength tester. The test method utilized the tapered beard from the length test (fibrograph-based length method), and the readings from the length test allowed for the clamps to be positioned along the length of the beard to ensure the proper mass of fiber was being broken (Naylor et al. 2014) (Fig. 3.17). This methodology eliminates the weighing of the beard; however there is no guarantee that the fibers being tested are continuous across both clamps. The original principles of the Motion Control Industries device have been carried through to the Uster HVI and most other SITCs.

The tapered beard approach uses a brushing apparatus to at least partially eliminate crimp. Maturity and fineness can affect estimated bundle mass and influence the measurement of tenacity. Therefore, micronaire is often used to provide a correction factor for the mass estimate. While the HVI system uses an estimate of bundle mass, the Fibrotest instrument manufactured by Textechno (Fig. 3.14) weighs the portion of the broken beard that is not retained in the fibrogram clamp in lieu of using the micronaire value to estimate mass.

3.4 Micronaire

Micronaire is an indirect measurement that has great importance in determining the market value of cotton but is of limited value to researchers in determining maturity and fineness. Micronaire is directly measured by compressing a known mass of

Fig. 3.17 Fiber beard after being broken for bundle tensile property assessment. The two small tufts on the outer edges of the beard were not clamped and broken during tensile property assessment

cotton to a known volume and measuring the pressure drop of a controlled airflow through the sample. Micronaire directly measures this relationship between the specific surface of the sample and airflow, a property that depends on fiber fineness and maturity. This relationship is governed by a theory of laminar fluid flow through particles known as Darcy's law. It was first related to textiles in 1927 (Kozeny 1927). Although micronaire was originally intended to measure the linear density of cotton and was reported with units of µg/inch (Lord 1955), it is now understood to be a unitless value and is reported on an arbitrary scale of relative values, which do not directly relate to any single physical property (Heap 2000).

Lord's equation (Eq. (3.1)) published in 1956 was based on a set of 100 cottons and provided a relationship between micronaire (Mic), maturity (M), and fineness (H). It is evident that the micronaire value alone is insufficient to determine either the maturity or fineness of a cotton sample:

$$MH = 3.86\text{Mic}^2 + 18.16\text{Mic} + 13.0 \tag{3.1}$$

Although micronaire is an airflow measurement, many instruments such as the now defunct Shirley F/MT (Shirley Developments Ltd., Manchester, England) and the relatively new Cottonscope (Cottonscope Pty Ltd., Ardross, Australia) (Fig. 3.18) report micronaire values which are derived from direct measurements of fiber fineness and maturity.

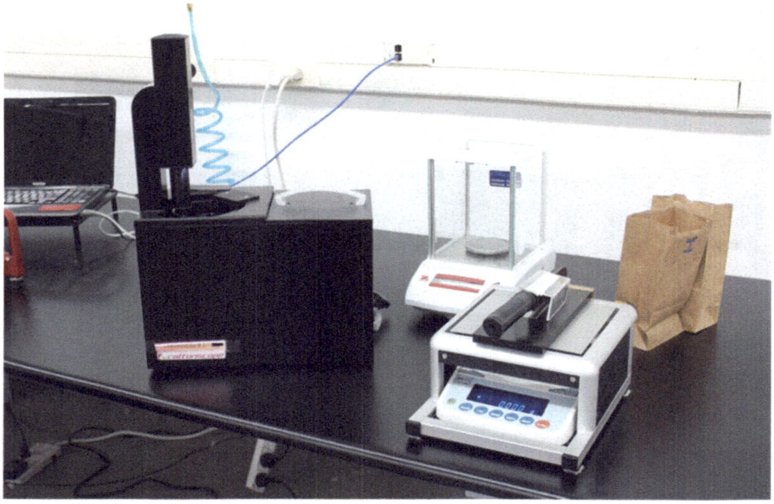

Fig. 3.18 Cottonscope

3.4.1 Micronaire Measurements

The direct measurement of micronaire is based on airflow. Most SITCs, such as the Uster HVI, utilize airflow to measure micronaire (Fig. 3.19). SITCs require a 10 g sample size for the measurement. Older instruments, such as the Fibronaire (Motion Control Industries, Dallas, TX) (Fig. 3.20), are able to conduct the same measurement using a smaller sample size of 3.2 g (50 grains). While the sample size of the Fibronaire can be of value to cotton breeders and other researchers working with small samples, the repeatability of the measurement is somewhat reduced due to the smaller sample size. Although Motion Control Industries is no longer in business, there are similar instruments produced around the world, and many of the original instruments are still in laboratory use.

Fiber preparation and calibration is important for airflow-based micronaire measurements. Packing cotton more tightly or loosely into the chamber can alter the test results, and calibration cottons are needed to reduce operator bias. Care should be taken to ensure there is minimal non-lint content in the sample as foreign matter will affect the airflow. Micronaire measurement of cotton is covered by two ASTM standards D1448 (Fibronaire) and D5867 (Cotton Classification Instruments) (ASTM D1448 2016; ASTM D5867 2016).

3.5 Fineness and Maturity

Cotton fiber fineness is defined as the linear density of the fiber or weight per unit length. Standard fineness is proportional to the diameter of the fiber and can be calculated by dividing fineness with maturity ratio. Cotton fibers are single

Fig. 3.19 Sample being ejected from an HVI micronaire chamber

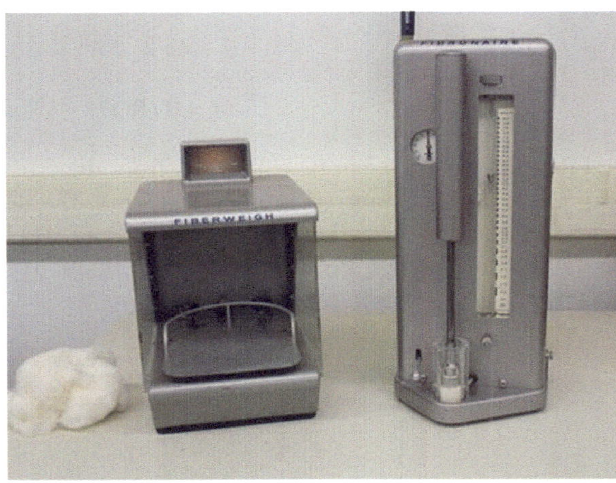

Fig. 3.20 FiberWeigh (50 grain scale) and Fibronaire testing apparatus. The Fibronaire instrument is loaded with a sample and providing a reading

elongated epidermal cells from the surface of a cotton seed whose diameter is established in the initial stages of fiber development (Stewart 1975). The fiber cell undergoes a stage of rapid elongation which is concluded around the time that deposition of cellulose in the secondary cell wall begins. As cotton fibers develop, cellulose is continually deposited creating the secondary cell wall in the inside of the cell making the lumen, the area where cell organelles reside, smaller and increasing the linear density of the fiber. Fiber maturity is less heritable than fineness because the development of the secondary cell wall is dependent on the environment (Bange et al. 2010). Fiber diameter plays an important role in textile processing as finer fibers allow more fibers within a cross section of yarn. A few techniques are available to measure cotton fiber fineness and maturity separately.

3.5.1 Cross-Sectional Image Analysis

The fundamental reference method for measuring both fiber fineness and maturity is to embed fiber bundles in a polymer resin, cross-section the resins, image the cross sections, and analyze the images for secondary cell wall area and fiber perimeter. The embedding and cross-sectional technique was described by Boylston et al. (1995) and has been used to create a set of 104 reference cottons for fiber maturity measurements (Hequet et al. 2006). An example of a cross-sectioned fiber bundle is shown in Fig. 3.21. The cell wall area and the perimeter of the fiber cell are the two independent parameters that define the morphology of a cotton fiber cross section (Thibodeaux and Rajasekaran 1999). The degree of thickening of the cell wall, θ, is also known as circularity and can be calculated from Eq. (3.2) where A_w is the cross-sectional area of the cell wall and P is the perimeter of the cell wall. The maturity ratio, M, can be calculated by Eq. (3.3) which was derived from sodium hydroxide swelling measurements (Pierce and Lord 1939):

$$\theta = 4\pi A_w / P^2 \tag{3.2}$$

$$M = \theta / 0.577 \tag{3.3}$$

Fiber fineness (H) can be calculated from the cross-sectional area and density (Eq. (3.4)) and is expressed in mtex (mg/km). The density (ρ) of the cell wall is generally stated as 1.52 g/cm^3. Fiber fineness can also be reported in terms of standard fineness, H_s, using Eq. (3.5) (Hequet et al. 2006):

$$H = \rho A_w \tag{3.4}$$

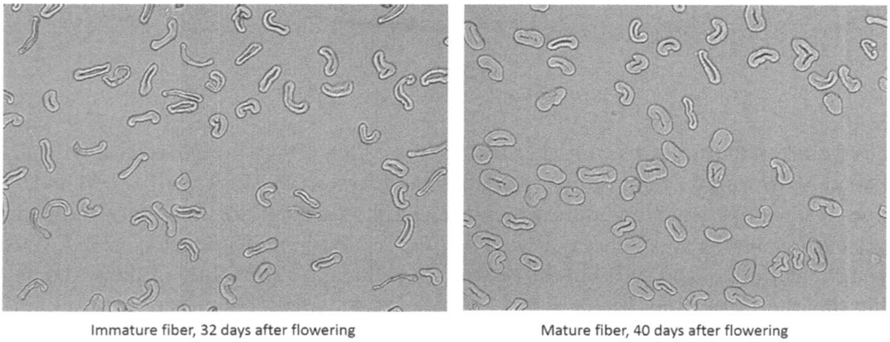

Fig. 3.21 Typical cross sections of cottons with differing maturity levels

$$H_s = \frac{H}{M} = \frac{0.577\rho}{4\pi} P^2 \tag{3.5}$$

Although fiber cross sections are the reference method, it is a tedious technique that cannot be applied to a large number of samples in a timely manner. Several hundred fiber cross sections are required to adequately describe the natural variation in fiber maturity and fineness within a given sample. Care must be taken to not bias the results through the removal of immature fibers during the preparation of the bundle for embedding. The image analysis software is not perfect and can incorrectly assess features. An operator must review each image after analysis to ensure the software correctly identified cell wall boundaries and that no air bubbles or other false readings are analyzed by the software.

3.5.2 AFIS

The Uster AFIS instrument provides measurements of both fineness and maturity. As previously described, a 0.5 g sample is hand-formed into a sliver and fed into a pinned cylinder to individualize the fibers. Individual fibers are suspended in an airstream where they pass through electro-optical sensors. Similar to the length measurement, these electro-optical sensors utilize two beams to measure the circularity (θ) of the individual fibers by comparing the measurements taken from two sensors oriented 40° to each other (Bradow et al. 1996). A flat fiber will have distinctly different "ribbon widths" when viewed by the separate sensors, while a circular fiber will have similar ribbon widths from the various perspectives. The AFIS calculates maturity ratio (M) using the ratio of the number of fibers with each θ value as shown in Eq. (3.6):

$$M = \frac{\theta \geq 0.5}{\theta \leq 0.25} \tag{3.6}$$

Similarly, the AFIS is able to report a fineness measurement for the fibers by applying a density calculation to the fibers measured for length and width. For both fineness and maturity, the AFIS provides an average value and a histogram (Figs. 3.22 and 3.23).

3.5.3 HVI

The Uster HVI 1000 is not able to measure maturity but it does report a related parameter called maturity index. The maturity index value is not part of cotton classing, and it is derived from proprietary equations based on micronaire readings and measurements taken during the length measurement process. HVI maturity

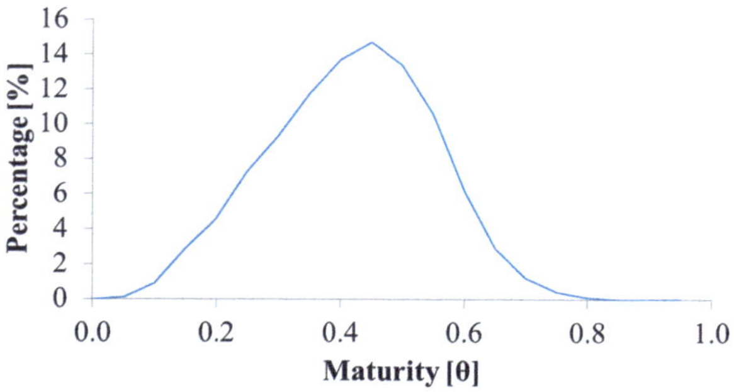

Fig. 3.22 Example of a maturity distribution reported by AFIS

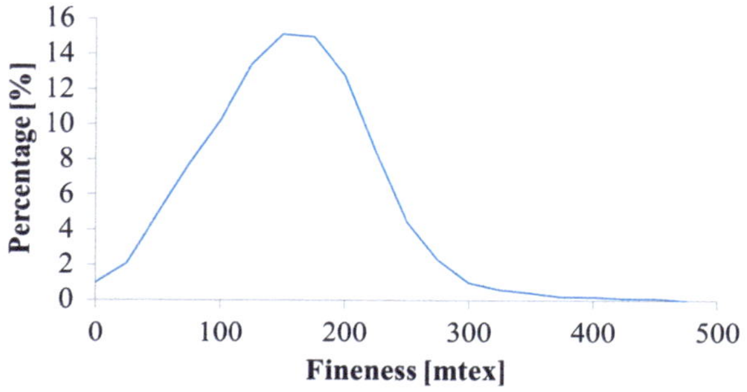

Fig. 3.23 Example of a fineness distribution reported by AFIS

index is an estimate and has limited value in germplasm development. In addition, there are many industry reports that the Uster HVI maturity index lacks the sensitivity needed to be useful for industrial applications. The Uster HVI does not report a separate fineness value.

3.5.4 *Cottonscope*

Recently, a new instrument has been developed that combines several existing techniques for fineness and maturity. This instrument, Cottonscope, uses automated image analysis and polarized light microscopy (ASTM D1442 2016, Procedure 1) for fiber maturity and fineness measurements. A 50 mg sample of fibers is cut into snippets using a double-bladed guillotine. The snippets are distributed in an agitated water bath and detected by a camera system which determines the physical parameters (Gordon et al. 2012).

3.5.5 Maturity Measurements

3.5.5.1 Polarized Light Method

Cotton fibers refract polarized light differently based on cell wall thickness. This allows fiber maturity to be inferred by the color shifts observed as fibers are rotated under a polarizing filter on a microscope (Meredith 1953). This method has largely been a manual method involving the examination of a small number of fibers (ASTM D1442 2016) although recently an instrument, SiroMat, was developed by CSIRO to automate the technique (Long et al. 2010). The SiroMat was later combined with another technique to form the basis of the Cottonscope (Fig. 3.24).

3.5.5.2 Differential Dyeing

Differential dyeing is a subjective measurement technique, which can be used to visually differentiate between mature and immature fibers. The technique is also referred to as the "Goldthwait Method" and makes use of two different colored dyes that have different molecular size. The procedure applies a mixture of the red and green dyes to raw cotton, and then the material is subjected to a vigorous rinsing to strip unadhered dye from the material. Immature cottons which are lacking in developed secondary cell wall are unable to accept the red dye leaving the immature fibers to show only the green dye. Mature cottons are able to hold the red dye and allow for a visual examination of the sample to determine if the sample is more green (less mature) or more red (more mature) (Goldthwait et al. 1947) (Fig. 3.25). Although not in common use due to the lack of quantitative results, this method does remain an ASTM standard practice, D1464 (ASTM D1464-12 2016).

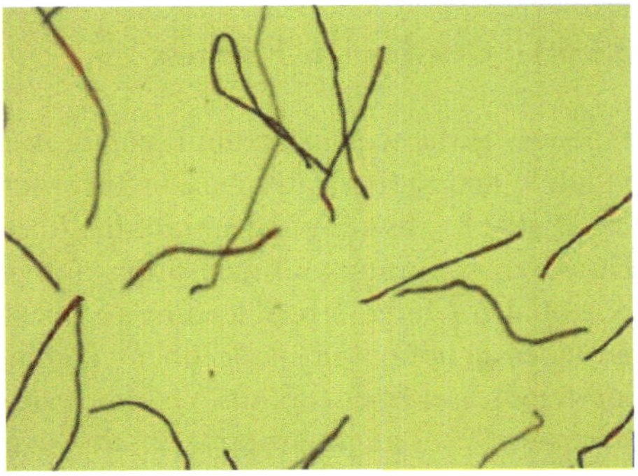

Fig. 3.24 Cottonscope instrument output showing the fiber snippet appearance due to birefringence of polarized light

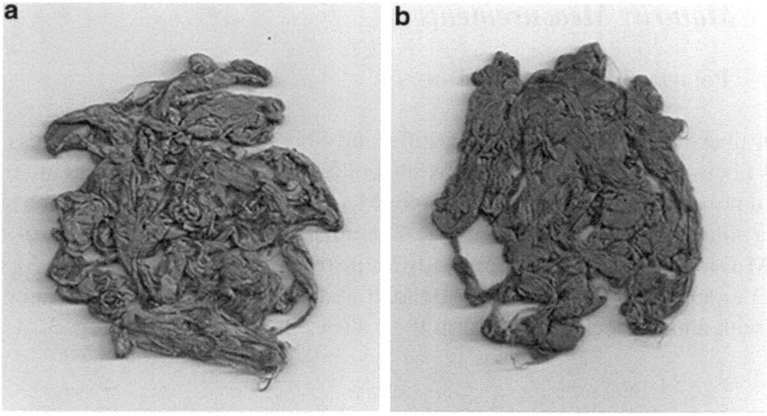

Fig. 3.25 Two cottons of differing maturity subjected to differential dyeing

3.5.5.3 Infrared (near Infrared (NIR) and Fourier Transform Infrared (FTIR))

Over the years there have been numerous researchers investigating the use of NIR and FTIR to measure cotton maturity (Ramey 1982; Montalvo and Von Hoven 2004; Abidi et al. 2008). These techniques can provide rapid and nondestructive analysis of cotton fiber. Although the calibration equations may vary based on the investigator and the specific technique used, these methods all rely on the measurement of chemical bonds and compounds formed during cellulose deposition which is the biological process of fiber maturation.

3.5.6 Other Fineness Measurements

3.5.6.1 Gravimetric Fineness

Fineness is the weight per unit length. A small number fibers can be weighed individually and individually measured for length allowing for the linear density to be calculated by dividing the sum of the fiber weights by the sum of the fiber lengths. However, weighing and measuring individual fibers is a tedious process with the potential for large errors leading to questionable accuracy. ASTM D1577 (2016) provides an option for single-fiber weighing (option B) but also an option for weighing bundles of fiber (option A). There are two approaches to cotton fineness using fiber bundles. Length groups prepared using the array method (ASTM D1440) can be weighed and an average length of the group applied to the number of fibers in the bundle. This method requires all of the fibers in each length group used to be counted. An alternative is to prepare a bundle of fibers such as those used in the flat bundle strength methods (ASTM D1445 2016). Once the bundle is loaded into the clamping device, the fibers extending outside of the clamping device are cut off resulting in the length of all of the fibers in the bundle being the length of the clamping device. These fibers may then be removed from the clamp, counted, and weighed.

3.5.6.2 Vibroscope

Fiber cross-sectional area can be measured by determining the natural resonant frequency of a cotton fiber held under tension. The frequency, tension, and gauge length of the fiber allow the cross-sectional area to be determined. The cross section together with the length of the fiber and density of the fiber can be used to estimate the average linear density of the fibers. This method was first developed in the 1940s by Gonsalves (1947). Although tedious, the technique was used for many years (Montgomery and Milloway 1952). The Favimat instrument by Textechno (Moenchengladbach, Germany) has automated the vibroscope method through the use of a variable frequency audio signal and an electronic observation of the fiber's oscillations. The determination of fineness using the vibroscope technique is option C in ASTM D1577 (2016).

3.6 Color

Color was one of the first physical properties of cotton used to assess the economic value and perceived quality of cotton. Upland cottons (*Gossypium hirsutum*) are various shades of white and off-white, while extra-long staple (ELS) cottons (*Gossypium barbadense*) are more creamy yellowish white colors. Initially, color was measured visually, and a number of subjective color grades were created to label the color of cotton. Today, color grading may be measured either visually or with instrumentation. In the United States, all cotton is classed using a colorimeter instrument as part of the HVI (Fig. 3.26). However, in many countries, the official color grade is still determined visually.

Color can be impacted by many factors, such as weather, insect activity, fungi, and moisture (Anthony 2002; Barker et al. 1979; Chun et al. 2006). The measurement of cotton color is generally a bulk measurement with the overall grade of the sample being measured. The visual grader utilizes the entire sample within their field of view, and instruments utilize the sample presented to the measuring window. Non-lint content, e.g., leaf and field trash, as well as the alignment of the fiber can alter the color grading results. The distribution of color within a cotton sample is not readily available; however, there have been some efforts to assess color distribution (Cui et al. 2014; Xu et al. 2001).

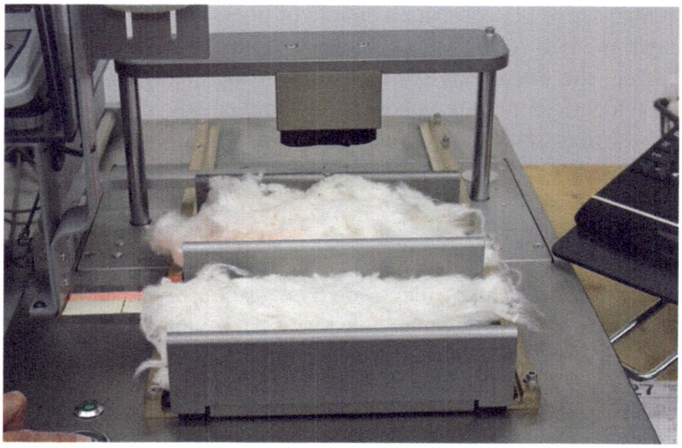

Fig. 3.26 Samples loaded into an HVI for color and trash analysis

3.6.1 Color Grades

The majority of cotton is assigned color grades using either the USDA Official Color Grades for American Upland Cotton (Fig. 3.27) or the USDA Color Grades for American Pima Cotton (Fig. 3.28). These color grades are used for both visual and instrument color determination. Although some countries do have their own color standards, many countries utilize the USDA grades. Upland cotton can be assessed into 30 different grades, while Pima cotton is divided into only six grades. The Upland color grades consist of two digit numbers with the first digit ranging between 1 and 8 and the second digit ranging between 1 and 5. The lower the first number the brighter the cotton appears, and the higher the second number the more yellow the cotton appears. Color measured by a colorimeter, either as part of an SITC or a stand-alone instrument, can be converted from the components of reflectance and yellowness into color grades using the official charts. Color measured visually is determined by physical cotton standards prepared annually by the USDA-AMS (Memphis, TN). Cotton color is known to change over time; therefore, physical standard references must be replaced on an annual basis.

3.6.2 Reflectance (Rd) and Yellowness (+b)

Color is three-dimensional in nature; however, cotton grading only considers two dimensions of color. Instrument-based color measurements utilized the percent reflectance (Rd) and the amount of yellowness (+b) in a sample of cotton. Work on instrument-based color measurement of cotton goes back more than 60 years to the development of the cotton colorimeter (Nickerson et al. 1950). Instrument color can

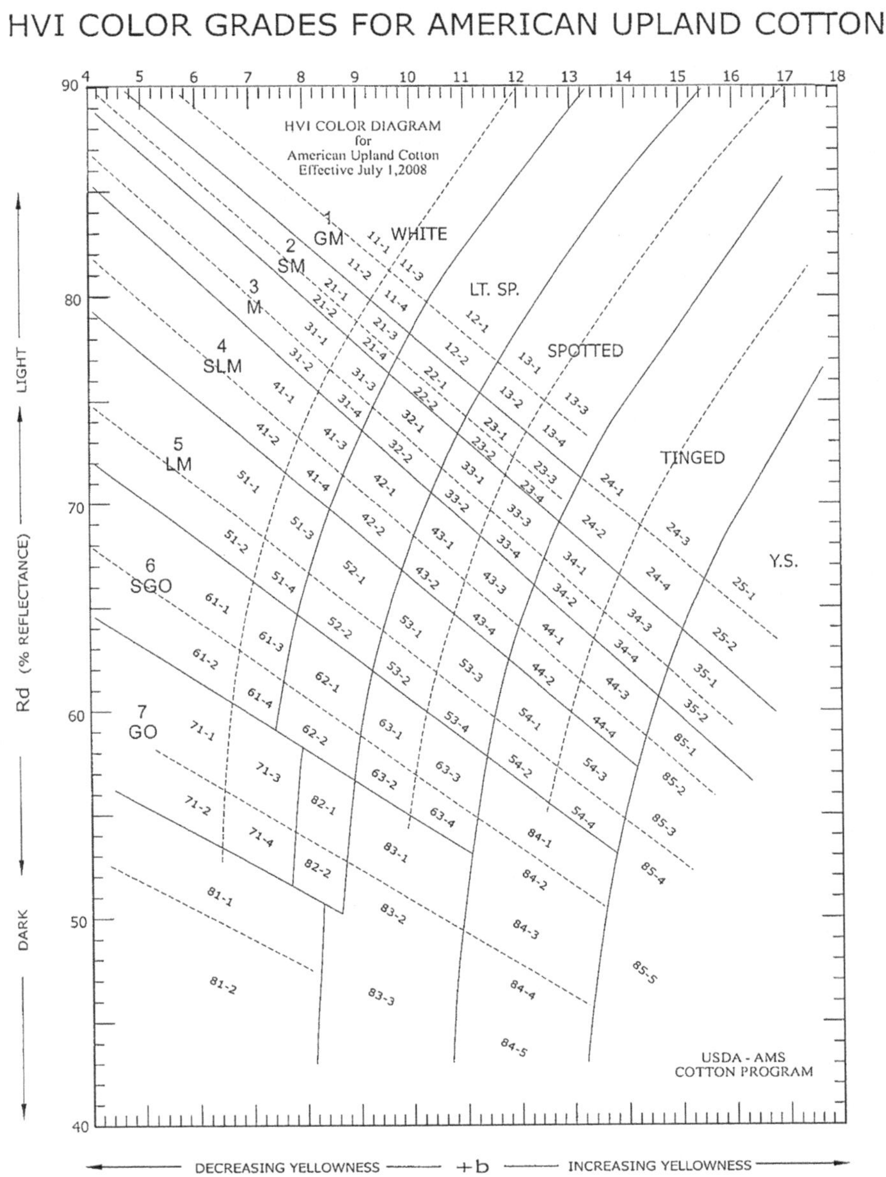

Fig. 3.27 USDA-AMS American upland cotton color grades

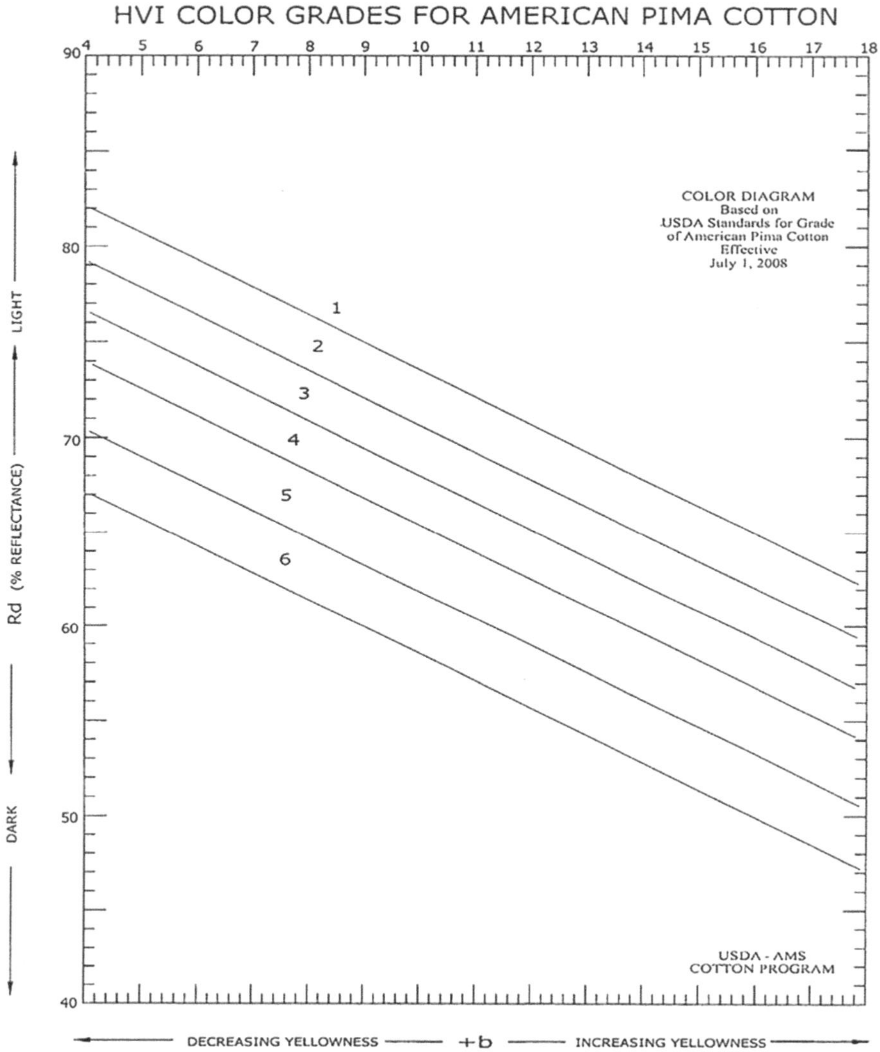

Fig. 3.28 USDA-AMS American pima cotton color grades

be reported in terms of the Rd. and +b values as well as the resultant color grade. Rd. and +b values can be converted to a color grade, but the reverse is not possible as a color grade consists of a large range of possible Rd. and +b values. There has been work performed to relate cotton color standards to more common color measurements such as CIELAB (Thibodeaux et al. 2008a). Measuring the color of cotton using typical laboratory colorimeters and spectrophotometers is possible; however care must be taken if the intent is to compare the results to cotton standard measurements (Rodgers et al. 2008). Researchers who are interested in statistical analysis and variation of color should use Rd. and +b values where possible statistical analysis cannot be easily performed on color grades.

3.7 Neps

Neps are defined as a small cluster of entangled fibers which may consist of only fibers or also contain foreign matter such as a seed coat fragment entrained within (van der Sluijs and Hunter 1999). Neps consist of a number of fibers with 16 or more being most common (Hebert et al. 1988). Pearson (1933) divided neps into 15 types; Lord (1948) divided neps into five categories; however, it is commonly accepted that neps can be classified into two broad groups of biological/raw and process/mechanical (Mehta et al. 1990). Biological/raw neps are generally found to either contain foreign matter, such as seed coat fragments, or to originate from aborted seed. Process/mechanical neps are composed mainly of fibers and are created during processing. Neps can be created due to handling of the seed cotton or during processing from fiber into yarn. Nep creation is heavily influenced by fiber maturity with less mature fibers being more likely to create neps during processing and handling (Anthony et al. 1988). Neps absorb less dye and reflect light in a different manner than the body of the yarn and therefore show up as a defect in finished goods; however, cotton classification does not measure neps, so there is no immediate penalty for cottons with high levels of neps. The reputation of producing high nep levels can damage a production region's reputation due to the impact on the spinning mill (Gordon et al. 2004). Testing for nep levels is important to researchers to understand both fiber maturity and the processing potential of a cotton. Neps are generally measured as a count per gram of cotton.

The measurement of neps may be done manually; however, this is uncommon, time-consuming, and subject to operator influence. The most common measurement technique for neps in cotton lint follows ASTM D5866 (2016). There are two commercial instruments most commonly used for the measurement of neps by this method, the Uster AFIS and the Neps and Trash Indicator (NATI) (Fig. 3.29) manufactured by Mesdan, SpA (Brescia, Italy). The principle behind automated nep testing is that a sliver is fed into an opening cylinder, which individualizes fibers into an airstream that carries them past an optical detector that counts the number of neps. Instrument manufacturers may report the total number of neps/gram of cotton, or they may differentiate the neps into size categories.

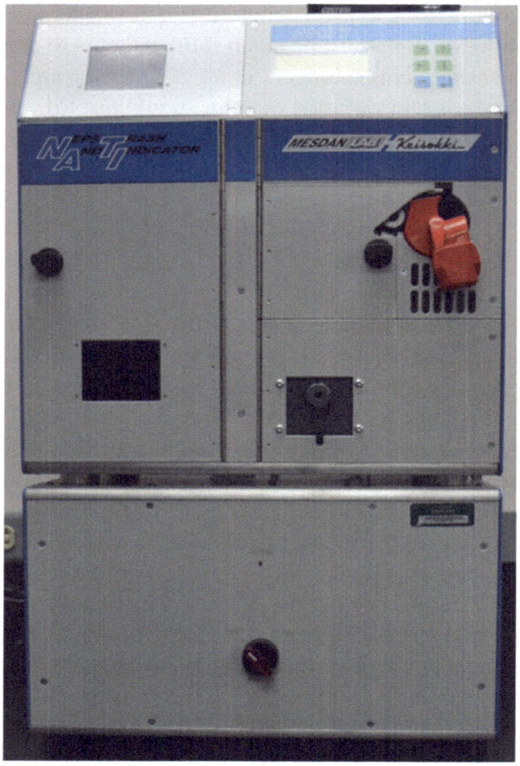

Fig. 3.29 Neps and trash indicator (NATI: by Mesdan SpA)

3.8 Fiber Cohesion

The process of converting fiber into yarn can be viewed as controlling the manner in which fibers slide past each other in carding, drawing, and spinning. Typical short-staple spun yarn is held together only by the surface friction of the fibers wrapped together. Fiber cohesion is the property that controls the friction between fibers and between fiber and machine components (Delhom et al. 2017). Although fiber cohesion is an important property of cotton fiber, it is not included in routine testing by SITC or most instruments. Fiber cohesion can be measured statically (ASTM D2612 2016) or dynamically (ASTM D4120 2016), although there are also some approaches to measuring cohesion forces through energy consumption during processing (El Mogahzy et al. 1998). All measurements of fiber cohesion require that fibers be aligned into a sliver or roving either mechanically or manually. Cohesion is affected by many fiber properties, including length, fineness, and crimp. However, these properties alone are not enough to calculate a value for fiber cohesion. As fibers become more aligned, the force required to break sliver decreases, so the amount of twist in slivers must be consistent to fairly compare results.

3.8.1 Static Cohesion

Static cohesion (ASTM D2612 2016) is measured using a constant-rate-of-extension tensile tester. A sliver or roving is mounted on a backing paper to prevent unintentional distortion of the fibers during sample loading. The gauge length is the nominal staple length plus 100 mm. The backing paper is cut once the sample is loaded and then the load to pull the fiber arrangement apart is measured. The specimen should be cut at the innermost portion of each clamp so that the mass of the tested specimen can be measured. The drafting tenacity can be calculated by multiplying the cohesive force measured by the test by the specimen length and dividing by the specimen mass.

3.8.2 Dynamic Cohesion

Dynamic cohesion (ASTM D4120 2016) is measured using two pairs of drafting rollers. The pairs of drafting rollers are moving at different speeds (i.e., the draft ratio) and should be designed to measure the force applied to the front pair of drafting rolls. The sole commercial source of the Fiber Cohesion Meter is Rothschild (Zurich, Switzerland); however the method is quite similar to the ITT Draftometer (Fig. 3.30) developed by the former Institute of Textile Technology (Charlottesville, VA) (Lin et al. 2011).

Fig. 3.30 Draftometer based on ITT design

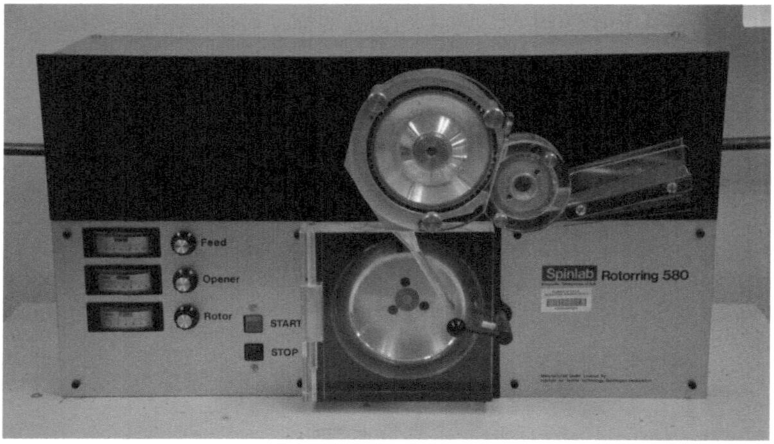

Fig. 3.31 Rotor Ring 580 as manufactured by Spinlab

3.8.3 Energy Consumption

The need to prepare uniform sliver or roving to perform the ASTM cohesion tests and the relatively small amounts of material that can be tested in this manner have led to the search for alternative techniques to measure cohesion. The concept of comparing the energy consumption required to process different cotton samples has gained popularity as an indirect method of fiber cohesion. Although any processing equipment can theoretically be used, the Rotor Ring instrument (Fig. 3.31) originally developed by the Institute of Textile Technology (Reutlingen, Germany) and later manufactured for commercial use by several manufacturers has been the focus of most research (El Mogahzy et al. 1998; Ghosh et al. 1992). Energy consumption by the Rotor Ring is a function of the fiber-to-fiber and fiber-to-metal friction and was found to be consistently repeatable for individual samples, which suggested that the energy differences between cottons is a function of friction. Measuring energy consumption using this testing approach results in a measurement expressed in joules, rather than a normalized force measurement often reported by other techniques.

3.9 Summary

Many physical parameters determine the quality of a cotton sample. These parameters are interrelated and can have an effect on the measurement of each other. The processing history of samples gathered for testing is an important consideration; for example, a hand harvested and ginned immature cotton may appear to have better fiber length than that same cotton subjected to mechanical harvesting and ginning during which fibers were broken and removed due to the processing. No single test can fully characterize a sample; different results can be obtained when different techniques are used to measure the same property, and material variability requires that replicate samples be taken to find a statistical representation of the true value.

There are many reasons cotton fiber quality is examined, not limited to marketing, quality control, and germplasm development. The processing need or research objective must be considered when selecting the appropriate fiber quality instrument. There are many test methods, and the proceeding discussion is an overview of the most common and influential properties and methods but is not an exhaustive or all-inclusive listing. New instruments and protocols are constantly being developed to keep pace with changing research and industry demands.

References

Abidi N, Hequet E, Cabrales L, Gannaway J, Wilkins T, Wells LW (2008) Evaluating cell wall structure and composition of developing cotton fibers using Fourier transform infrared spectroscopy and thermogravimetric analysis. J Appl Polym Sci 107:476–486

Anthony WS, Meredith WR, Williford JR (1988) Neps in ginned lint: the effect of varieties, harvesting, and ginning practices. Text Res J 58:633–639

Anthony WS (2002) Impact of moisture added at lint slide on cotton color, vol 103. Cotton Gin and Oil Mill Press, Mesquite, pp 8–12

ASTM D1440-07 (2012) Standard test method for length and length distribution of cotton Fibers (Array method). ASTM International, West Conshohocken

ASTM D1442-06 (2016) Standard test method for maturity of cotton Fibers (sodium hydroxide swelling and polarized light procedures). ASTM International, West Conshohocken

ASTM D1445/1445M-12 (2016) Standard test method for breaking strength and elongation of cotton Fibers (flat bundle method). ASTM International, West Conshohocken

ASTM D1448-11 (2016) Standard test method for micronaire reading of cotton fibers. ASTM International, West Conshohocken

ASTM D1464-12 (2016) Standard practice for differential dyeing behavior of cotton. ASTM International, West Conshohocken

ASTM D1577-07 (2016) Standard test methods for linear density of textile fibers. ASTM International, West Conshohocken

ASTM D1776/D1776M-16 (2016) Standard practice for conditioning and testing textiles. ASTM International, West Conshohocken

ASTM D2612-99 (2016) Standard test method for fiber cohesion in sliver and top (static tests). ASTM International, West Conshohocken

ASTM D4120-01 (2016) Standard test method for fiber cohesion in roving, sliver, and top in dynamic tests. ASTM International. West, Conshohocken

ASTM D5866-12 (2016) Standard test method for Neps in cotton fibers. ASTM International, West Conshohocken

ASTM D5867-12e1 (2016) Standard test method for measurement of physical properties of raw cotton by cotton classification instruments. ASTM International, West Conshohocken

Bange MP, Long RL, Constable GA, Gordon SG (2010) Minimizing immature fiber and neps in upland cotton. Agron J 102:781–789

Bargeron JD (1986) Preliminary investigation of the length measurement of cotton fibers with the Peyer Texlab system: compatibility and repeatability. Text Res J 56:121–123

Barker GL, McClendon RW, Colwick RF, Jones JW (1979) Relationship between cotton lint color and weather exposure. Trans ASAE 22:470–474

Boylston EK, Evans JP, Thibodeaux DP (1995) A quick embedding method for light microscopy and image analysis of cotton fibers. Biotech Histochem 70:24–27

Bradow JM, Hinojosa O, Wartelle LH, Davidonis G, Sassenrath-Cole GF, Bauer PJ (1996) Applications of AFIS fineness and maturity module and X-ray fluorescence spectroscopy in fiber maturity evaluation. Text Res J 66:545–554

Bragg CK, Shofner FM (1993) A rapid, direct measurement of short fiber content. Text Res J 63:171–176

Cai Y, Cui X, Rodgers J, Thibodeaux D, Martin V, Watson M, Pang S (2013) A comparative study of the effects of cotton fiber length parameters on modeling yarn properties. Text Res J 83:961–970

Chun DTW, McAlister DD, Hughs SE, Cobb DR (2006) Microbial census and evidence for a direct temporal effect of bale moisture on color grade during six months of storage. J Cotton Sci 10:201–209

Cui X, Calamari TA, Suh M (1998) Theoretical and practical aspects of fiber length comparisons of various cottons. Text Res J 68:467–472

Cui X, Cai Y, Rodgers JE, Martin VB, Watson MD (2014) An investigation into the intra-sample variation in the color of cotton using image analysis. Text Res J 84:214–222

Delhom CD, Martin VB, Schreiner MK (2017) Textile industry needs. J Cotton Sci 21:210–219

DIN 53806:1970-02 (1970) Testing of textiles; length determination for cotton with a comb sorter. DIN Deutsches Institut für Normung e.V., Berlin

El Mogahzy Y, Broughton R, Guo H, Taylor RA (1998) Evaluating staple fiber processing propensity part 1: processing propensity of cotton fibers. Text Res J 68:835–840

Foulk JA, McAlister DD (2002) Single cotton fiber properties of low, ideal and high micronaire values. Text Res J 72:885–891

Frydrych I (1995) Relation of single fiber and bundle strengths of cotton. Text Res J 65:513–517

GB/T6098.1-1985 (1985) Test method of cotton fiber length using roller analyzer. Chinese Cotton Standard

Ghosh S, Rodgers JE, Ortega AE (1992) Rotor ring measurement of fiber cohesion and bulk properties of staple fibers. Text Res J 62:608–613

Goldthwait CF, Smith HO, Barnett MP (1947) New dye technique shows maturity of cotton. Textile World July:105–110

Gonsalves VE (1947) Determination of denier and strength of single filaments by vibroscope and Heim tensile tester. Text Res J 17:369–375

Gordon SG, van der Sluijs MHJ, Prins MW (2004) Quality issues for Australian cotton from a mill perspective. Australian Cotton Cooperative Research Centre, Narrabri

Gordon S, Naylor G, Brims MA (2012) Cottonscope: a new instrument for maturity and fineness measurements. In: Proceedings of the 31st international cotton conference, March 2012, pp. 21–24

Heap SA (2000). The meaning of micronaire. In: Proceedings of Bremen Cotton Conference, Bremen, Germany, p. 17

Hebert JJ (1975) Effect of convolution angle upon fiber strength. Text Res J 45:356–357

Hebert JJ, Boylston EK, Thibodeaux DP (1988) Anatomy of a nep. Text Res J 58:380–382

Hebert JJ, Thibodeaux DP, Shofner FM, Singletary JK, Patelke DB (1995) A new single fiber tensile tester. Text Res J 65:440–444

Hequet EF, Wyatt B, Abidi N, Thibodeaux DP (2006) Creation of a set of reference material for cotton fiber maturity measurements. Text Res J 76:576–586

Hertel KL (1940) A method of fibre-length analysis using the fibrograph. Text Res J 10:510–525

Kozeny IJ (1927) Über Grundwasserbewegung. Wasserkraft und wasserwirtschaft 22:67

Lin Q, Oxenham W, Yu C (2011) A study of the drafting force in roller drafting and its influence on sliver irregularity. J Text Inst 102:994–1001

Liu J, Hongbo Y, Hsieh Y (2001) Variations of mature cotton fiber tensile properties: association with seed position and fiber length. Text Res J 71:1079–1086

Long RL, Bange MP, Gordon SG, Constable GA (2010) Measuring the maturity of developing cotton fibers using an automated polarized light microscopy technique. Text Res J 80:463–471
Lord E (1948) Neppy cotton: origin and cure. Emp Cotton Grow 25:180–190
Lord E (1955) Airflow through plugs of textile fibres, part 1 – general flow relations. J Text Inst 46:191–213
Lord E (1956) Airflow through plugs of textile fibres, part 2 – the micronaire test. J Text Inst 47:16–47
Mehta RD, Salame PA, Combs RN (1990) Dyeing of immature cotton neps covered with a cationic polymer. American Dyestuff Rep 79:38
Meredith R (1953) Measurements of orientation in cotton fibres using polarized light. Br J Appl Phys 4(12):369
Montalvo JG, Von Hoven T (2004) Analysis of cotton. In: Near-infrared spectroscopy in agriculture. Agronomy monograph no. 44. Crop Science Society of America, Madison
Montgomery DJ, Milloway WT (1952) The vibroscopic method for determination of fiber cross-sectional area. Text Res J 22:729–735
Naylor GR, Delhom CD, Cui X, Gourlot JP, Rodgers J (2014) Understanding the influence of fiber length on the High Volume Instrument™ measurement of cotton fiber strength. Text Res J 84:979–988
Nickerson D, Hunter RS, Powell MG (1950) New automatic colorimeter for cotton. J Optic Soc Am 40:446–449
Orr RS, Weiss LC, Grant JN (1955) The relation of single-fiber to flat-bundle strength and elongation of cotton. Text Res J 25:939–946
Pearson NL (1933) Neps and similar imperfections in cotton. USDA Tech Bull 396:23
Pierce FT, Lord E (1939) The fineness and maturity of cotton. J Text Inst 30:173–210
Ramey HH (1982) Estimating quality components of natural fibers by near-infrared reflectance: part 1: cotton fiber cross-sectional area and specific surface. Text Res J 52:20–25
Rodgers JE, Thibodeaux DP, Cui X, Martin VB, Watson MD (2008) Instrumental and operational impacts on spectrophotometer color measurements. J Cotton Sci 12:287–297
Rouse JT (1964) Cotton fiber strength tests at 1/8-inch gauge with pressley and stelometer instruments. Text Res J 34:908–910
Sasser PE, Shofner FM, Chu YT, Shofner CK, Townes MG (1991) Interpretations of single fiber, bundle, and yarn tenacity data. Text Res J 61:681–690
Stewart JM (1975) Fiber initiation on the cotton ovule (*Gossypium hirsutum*). Am J Bot 62:723–730
Thibodeaux DP, Rajasekaran K (1999) Development of new reference standards for cotton fiber maturity. J Cotton Sci 3:188–193
Thibodeaux DP, Rodgers JE, Campbell JH, Knowlton J (2008a) The feasibility of relating HVI color standards to CIELAB coordinates. AATCC Rev 8:44–48
Thibodeaux DP, Senter J, Knowton JL, McAlister D, Cui X (2008b) The impact of short fiber content on the quality of cotton ring spun yarn. J Cotton Sci 12:368–377
Urquhart AR, Eckersall N (2008) 32 – the moisture relations of cotton. Vii – a study of hysteresis. J Text Inst Trans 21(10):T499–T510
van der Sluijs MHJ, Hunter L (1999) Neps in cotton lint. Text Prog 28(4):1–47
Xu B, Huang Y, Watson MD (2001) Cotton color distributions in the CIE L*a*b* system. Text Res J 71:1010–1015
Zeidman MI, Batra SK, Sasser PE (1991) Determining short fiber content in cotton part 1: some theoretical fundamentals. Text Res J 61:21–30

Chapter 4
Chemical Composition and Characterization of Cotton Fibers

Yongliang Liu

4.1 Introduction

Cotton fiber is the dried cell walls of formerly living cells. Its formation originates from an ovary of the cotton flower and grows into a mature seed-containing cotton boll (or fruit) within approximately a 1.5- to 2-month growth period (Kim 2015; Stiff and Haigler 2012; Wakelyn et al. 2007a). Fiber development is considered to include four overlapping but distinctive stages: initiation, primary cell wall (PCW) formation for fiber elongation, secondary cell wall (SCW) biosynthesis for cellulose deposition and cell wall thickening, and maturation.

Fiber initiation commonly starts on or slightly before the day of anthesis and continues to at least 5 days post anthesis (DPA). The number of fiber initials per ovule or fiber initial density was observed to be correlated positively with cotton fiber yield (Lee et al. 2007; Li et al. 2009), but there was no correlation between fiber initial density and lint percentage in one report (Romano et al. 2011).

Following the initiation, fiber cells elongate rapidly and longitudinally until 20 DPA, reaching the fiber final length of 22–35 mm and depositing a thin PCW layer. During this process, fibers develop as twisted bundles and only become isolated from each other at the end of the fiber elongation stage. Elongating fibers show apparent morphological and physiological variations, which are controlled by genes, kinases and enzymes, and other plant regulators (Kim 2015; Stiff and Haigler 2012; Wakelyn et al. 2007a). Proteins, pectins, fatty acids, calcium ion, and sugars are related closely to the PCW biosynthesis process. About 22% semicrystalline cellulose fibrils held in a matrix of other polysaccharides were reported (Meinert and Delmer 1977; Singh et al. 2009).

Y. Liu (✉)
Cotton Structure and Quality Research Unit, USDA-ARS, Southern Regional Research Center, New Orleans, LA, USA
e-mail: yongliang.liu@ars.usda.gov

D. D. Fang (ed.), *Cotton Fiber: Physics, Chemistry and Biology*,
https://doi.org/10.1007/978-3-030-00871-0_4

Overlapping with final stage of PCW biosynthesis, the SCW synthesis starts from 15 to 22 DPA and continues for an additional 30–40 days. The cellulose increases quickly during this transition, and the deposition has been verified by conventional microscopy and other instrumental methods (Abidi et al. 2008; Hsieh 2007; Kim 2015; Kim et al. 2014; Liu 2015; Thibodeaux and Evans 1986; Wakelyn et al. 2007b). The thickened SCW is composed of pure cellulose microfibrils in a very ordered arrangement, and these microfibrils are oriented helically along the growing fiber and change or reverse their directions periodically. Ordered or crystalline cellulose fractions have been investigated primarily by traditional X-ray diffraction (XRD) pattern (Hsieh et al. 1997). The cellular, biochemical, and genetic regulation of the SCW synthesis is the main interest of ongoing studies in cotton fiber biology (Fang and Percy 2015; Kim 2015; Stiff and Haigler 2012). As cellulose is a linear β 1 $\rightarrow$ 4-linked glucose residue produced by the photosynthesis process, environment factors (e.g., water, temperature, and nutrient stress) play important roles in modulating SCW formation.

At maturation (~40 to 45 DPA), cotton bolls dehisce, and fibers are fully exposed to air and sunlight. Accompanied by dehydrating, fibers collapse and change into flattened and twisted ribbon structures in cylindrical-like shapes. The cross-sectional image analysis clearly shows that mature fibers contain thicker SCW area than immature fibers.

The term DPA, frequently used to describe chronological fiber developmental period between floral anthesis and its harvest, cannot be used as a primary parameter to monitor the degree of fiber maturation (Bradow and Davidonis 2000). In cotton industry, fiber maturity has been accepted as reflecting the degree of the secondary cell wall thickening relative to the diameter of the fiber. Current-in-use instrumental methods of determining fiber maturity include cross-sectional image analysis, a direct but slow protocol, and standardized high volume instrument (HVI™), an indirect but much rapid test. HVI micronaire is one of the most essential cotton fiber characteristics and represents fiber maturity (degree of secondary cell wall development) and fineness (weight per unit length) simultaneously (Lord 1956). Hence, it has been increasingly and routinely utilized in the cotton and textile industry from cotton breeding program to textile quality control (Frydrych and Thibodeaux 2010; Kelly et al. 2012; Kim et al. 2014; Paudel et al. 2013).

Cellulose (88.0–96.5%) is a major chemical component in mature fibers, followed by noncellulosic constituents such as proteins (1.0–1.9%), waxes (0.4–1.2%), pectins (0.4–1.2%), inorganics (0.7–1.6%), and other substances (0.5–8.0%) (Hsieh 2007; Wakelyn et al. 2007c). Their contents change with cotton cultivar, growing environment (soil, water, temperature, farming practice, etc.), and the degree of fiber maturity. Mature fibers contain more cellulose and less noncellulosic components than immature fibers. Pure cellulose is located in the SCW area, whereas the noncellulosic constitutes coexist either on the outer layers (cuticle and PCW) or inside the lumens of the fibers. The presence of large amounts of immature fiber in commercial cottons has been found to cause fiber entanglement and form neps during mechanical processing and to degrade the desired color appearance in dyed yarn and finished fabric products.

4.2 Physical Characterization of Cotton Fibers

Dried and mature fibers are arranged in multilayer helical structures of cuticle, PCW, SCW, and the lumen from outside to inside (Hearle 2007; Kim 2015; Wakelyn et al. 2007b). There exists a winding or transition layer between the PCW area and the SCW area. The SCW area contains additional sub-layers that impact the wall thickness and further fiber maturity and fineness. In mature fibers, SCW cellulose dominates the fiber mass and consists of helical layers of nearly parallel microfibrils assembled in a spiral arrangement. The direction of the spiral along the fiber axis reverses at varying intervals randomly within the fiber, and these reversals could be related to breaking strength in cotton fibers (Hearle and Sparrow 1971).

Cotton fibers shrink and collapse with fiber drying, resulting in twisted and convoluted cotton fibers with hollow surface, thus affecting fiber cohesion, friction, and strength (Cho et al. 1996; Foulk and Mcalister 2002). The dried fibers appear in irregular and kidney-bean shape at times when viewing their cross sections.

4.3 Direct and Nondestructive Analysis of Cellulosic and Noncellulosic Components

Cellulose and noncellulosic components in cotton fibers have been analyzed through chemical extraction and separation steps routinely and regularly. As cotton fiber cellulose is not easily dissolved in most solvents, the determination of cellulose and noncellulosic components by conventional methods experiences significant drawbacks that include the tedious procedures of optimal solvent and temperature selection as well as extracted specimen identification.

As the need for rapid, direct, and nondestructive methods to identify both cellulose and noncellulosic components in cotton fibers, several state-of-the-art analytical techniques have been explored, including attenuated total reflection Fourier transform infrared (ATR FT-IR) spectroscopy, differential scanning calorimeter (DSC), thermogravimetric analysis (TGA), and pyrolysis-gas chromatography/mass spectroscopy (GC/MS) methods (Abidi and Manike 2017; Abidi et al. 2008, 2010, 2014; Hartzell-Lawson and Hsieh 2000; Liu et al. 2011). These rapid and nondestructive measurements have provided clear evidence of various noncellulosic components in developing cotton fibers through the onset of SCW synthesis. Waxy compounds in developing fibers up to 17 days are evident by their melting endotherms in TGA thermograms (Abidi et al. 2014). The presence of proteins in developing fibers up to 22 DPA can be monitored by the unique bands at 1543 cm^{-1} in ATR FT-IR spectra (Abidi et al. 2010, 2014). Pectins are detected by ATR FT-IR measurement in the 14 DPA and mature fibers. The presence of proteins in 14 DPA fibers is also detected by pyrolysis-GC/MS analysis (Hartzell-Lawson and Hsieh 2000). Among the analytical techniques examined, only pyrolysis-GC/MS can detect the presence of noncellulosic compounds in 27 DPA fibers (Hartzell-Lawson

Fig. 4.1 One type of ATR sampling device

and Hsieh 2000). In general, ATR FT-IR spectroscopy has evolved as an important alternative to examine cotton fiber development. It requires minimal sample preparation by ATR sampling on a small bundle of cotton fibers as little as 0.5 mg, permits routine analysis rapidly and nondestructively, and is easy to operate (Abidi and Manike 2017; Abidi et al. 2008, 2010, 2014; Lee et al. 2015; Liu and Kim 2015; Liu et al. 2011, 2012; Santiago and Hinchliffe 2015). One type of ATR sampling device is shown in Fig. 4.1 (Liu and Kim 2015; Liu et al. 2011, 2012). It has a 2 mm diameter sampling window and utilizes a DuraSamplIR single-pass diamond-coated internal reflection accessory. It enables the rapid and nondestructive characterization of small amounts of cotton fibers grown in vitro by cotton genomic researchers at times, although cotton fiber is easily available in large quantities.

4.4 Determination and Characterization of Cotton Fiber Cellulose

There is an abundant literature available on the biosynthesis of cotton fiber cellulose and its chemical and physical structure (Gordon and Abidi 2017; Gordon and Hsieh 2007; Kim 2015; Stiff and Haigler 2012; Wakelyn et al. 2007). Compared to traditional chemical analysis, direct and indirect instrumental methods have been developed to characterize the fiber cellulose quantitatively or qualitatively in a rapid and nondestructive way (Liu 2015). The direct methods include the well-established cross-sectional image analysis, the scanning electron microscope (SEM), and ATR FT-IR spectroscopy, while indirect instrument methods include HVI lines, the advanced fiber information system (AFIS), and the Cottonscope®.

4.4.1 Chemical Analysis of Fiber Cellulose Formation

Gel permeation chromatography (GPC) shows that cell wall polymers from fibers at the PCW stage have lower molecular weight than do the cellulose from fibers at the SCW stage (Timpa and Triplett 1993). XRD and microscopy chemical data indicate lower molecular weight, crystalline size, and extent of crystallinity within the PCW of the cotton fiber (Boylston and Hebert 1995). Both the anthrone method (Viles and Silverman 1949) and the Updegraff method (Updegraff 1969) have been used routinely to determine fiber cellulose content with the use of a cellulose standard, taking at least 2 days to measure cellulose content for each sample mostly due to slow hydrolysis of fiber cellulose.

The cell wall composition of developmental fibers has been analyzed from the early stages of elongation through the period of the SCW formation by extraction, separation, and chemical and instrumental determination. During the elongation period of *Gossypium* (*G.*) *hirsutum* L. Acala SJ-1 fibers, the levels of cellulose, neutral sugars (rhamnose, fucose, arabinose, mannose, galactose, and noncellulosic glucose), uronic acids, and total protein were observed to undergo marked changes (Meinert and Delmer 1977). The results suggest that the thickness of the cell wall is relatively constant until ~12 DPA and then increases rapidly until the SCW deposition is completed. Increases in all components contribute to the total wall increase between 12 and 16 DPA. The deposition of the SCW cellulose begins at about 16 DPA (at least 5 days prior to the cessation of elongation) and continues until about 32 DPA, accompanied by a quick decrease in protein and uronic acid content. Some of neutral sugars change in their contents during development, but the most dominant change is a large increase in noncellulosic glucose, which occurs just prior to the onset of the SCW cellulose deposition. Unlike the neutral sugars, cell wall amino acid composition does not change greatly during fiber development.

The amount of protein in *G. arboreum* L. fiber cell walls from one ovule was observed to reach a maximum value at the end of the elongation period, and it then decreases, before reaching a second maximum at the end of the SCW deposition (Huwyler et al. 1979). The amount of fucose, galactose, mannose, rhamnose, arabinose, uronic acid, and noncellulosic glucose reaches a maximum at the end of PCW formation or at the beginning of SCW deposition. The amounts of xylose and cellulosic glucose increase until the end of the fiber development. The amounts of glucose and arabinose decrease during SCW formation, indicating a conversion of some hemicellulosic components into cellulosic material.

Abidi et al. (2010) compared the changes in the amounts of cellulose and four major sugars (i.e., sucrose, glucose, fructose, and galacturonic acid) between two developmental fibers (*G. hirsutum* L. cv TX19 and TX55). The percentages of the four sugars decrease significantly for the two cultivars during the fiber development, whereas the amount of cellulose increases as a result of SCW development. Noticeably, TX19 fibers have a rapid cellulose synthesis from 10.7% to 80.4% between 14 and 24 DPA, with a clear increase at 18 DPA (56.9%). Relatively, cellulose content in developmental TX55 fibers was 9.1% between 10 and 20 DPA, rising to 34.1% at 21 DPA and to 68.3% at 24 DPA, before staying nearly unchanged from 27 to 56 DPA. The authors postulated that the earlier start of cellulose synthesis in TX19 fibers (18 DPA) compared with the TX55 cultivar (24 DPA) could have a significant impact on fiber maturity at the end of the growing season. They attributed the 6-day difference to more elevated enzymatic activities in the TX19 fibers than in the TX55 fibers.

4.4.2 ATR FT-IR Characterization of Fiber Cellulose Formation

As a complementary approach to cellulose and sugar content analysis, Abidi et al. (2010) applied ATR FT-IR spectroscopy to investigate the compositional and structural changes between two cultivars (TX19 *vs.* TX55). They analyzed a number of unique IR bands and related the DPA-dependent IR intensity variations with cotton cellulose formation. In particular, the intensity of the 710 cm^{-1} band (CH_2 rocking vibration in cellulose I_β) increases linearly between 10 and 30 DPA for the TX19 fibers, while a major intensity change of the 710 cm^{-1} band happens only at 21 DPA among TX55 fibers. Further, principal component analysis (PCA) identifies two groups of spectra (or samples) for each cultivar and highlights the difference between two cultivars. For TX19 fibers, Group 1 includes the spectra of fibers at 10, 14, and 17 DPA with negative PC1 (the first principal component) scores, and Group 2 includes the spectra of fibers from 18 to 56 DPA with positive PC1 scores. For TX55 fibers, Group 1 represents the spectra of fibers from 10 to 21 DPA with negative PC1 scores, and Group 2 includes the spectra of fibers from 24 to 56 DPA, possessing positive PC1 scores. Therefore, they concluded different transition phases between two cultivars, that is, the transition occurs from 17 to 18 DPA for TX19 fibers and between 21 and 24 DPA for TX55 fibers. The finding was in good agreement with chemical measurements of sugar and cellulose contents from previously established analytical protocols (Abidi et al. 2010).

Abidi et al. (2014) used similar strategy to compare the fiber development in two cotton cultivars (Texas Marker-1 (TM-1) and TX55) from 10 to 56 DPA. They suggested that ATR FT-IR spectral intensity changes at 3286, 1738, 1639, 1543, 1161, 897, and 710, and 667 cm^{-1} are useful indicators to monitor the deposition of the SCW cellulose. The intensities of two bands at 667 and 897 cm^{-1} were observed to be correlated with percentage of cellulose determined by the anthrone method (Viles and Silverman 1949), providing a relatively easy means of estimating cotton fiber cellulose content indirectly.

Recently, Liu and Kim (2017) linked the cellulose content determined by chemical analysis with ATR FT-IR spectral indices acquired by the reported procedures among developmental TM-1 and immature fiber (*im*) mutant cotton fibers. They found that the *R* value, infrared crystallinity (CI_{IR}), and the integrated intensity of the 895 cm^{-1} band exhibit strong and linear relationships with cellulose content.

4.4.3 ATR FT-IR Determination of Cotton Fiber Cellulose Maturity

Liu et al. (2011) developed two simple algorithms (referred to as R_1 and R_2) for the ATR FT-IR spectral discrimination between immature and mature fibers. The R_1 values were observed to increase with R_2 values for a dataset of 402 fibers. Setting an R_1 threshold value at 0.40, 197 of 201 (98.0%) immature fibers and 190 of 201 (94.5%) mature samples were correctly identified. With an R_2 threshold value at 2.24, only six immature fibers and ten mature samples were misidentified, producing an overall 96% accuracy in their correct differentiation. Next, they proposed a formula to estimate the degree of cotton fiber cellulose maturity (M_{IR}) by representing the R_1 values. In the concept, the M_{IR} values of 0.0 and 1.0 were assigned to the most immature and mature fibers, respectively. Therefore, immature fibers whose $R_1 < 0.40$ correspond to a $M_{IR} < 0.58$ in the maturity range of 0–1.0 and vice versa. Furthermore, they used cotton fibers with various maturity readings as determined from traditional image analysis (IA) and AFIS to validate the efficiency of accessing the M_{IR} from direct ATR FT-IR measurement. Strong correlations between M_{IR} against referenced IA and AFIS maturity readings were reported within small and selected sample sets.

In addition to the two simple algorithms (R_1 and R_2), Liu and Kim (2015) proposed third ATR FT-IR algorithm (R values) to compare cotton fiber development between *in planta* and in culture. Just like PCA, three simple algorithms involving the IR bands in the 1500–700 cm^{-1} region are capable of revealing the phase transition from the PCW to SCW biosyntheses and also of monitoring the phase transition between two types of fibers. In particular, they noted that the algorithms might be more effective than the PCA in detecting the change in fiber development. Compared to PCA result in Fig. 4.2a, both infrared crystallinity index (CI_{IR}) values in Fig. 4.2b and the R values in Fig. 4.2c suggest a clear difference between the developing fibers (>25 DPA) grown in planta and those in culture (Liu and Kim 2015).

4.4.4 Assessing the Cotton Fiber Cellulose Crystallinity

The term crystallinity index (CI) has been used to describe the relative amount of crystalline portion in a simple two-phase (crystalline vs. amorphous) cellulose model from XRD measurement. Two different XRD pattern analyses have been taken to estimate the CI of cotton fiber cellulose. The first one is the Segal method (Segal et al. 1962), which is two XRD peaks ($2\theta = 22.7°$ and 18°) based on simple formula. The second approach is a deconvolution method, in which individual crystalline peaks were extracted by a curve-fitting process from the diffraction intensity profiles (Hindeleh 1980; Hsieh et al. 1997, 2000; Hu and Hsieh 2001; Park et al. 2010).

Resolving XRD pattern of developing cotton fibers into at least four peaks located at $2\theta = 14.7$, 16.6, 22.7, and 34.4° that are indicative of the 101, $10\bar{1}$, 002, and

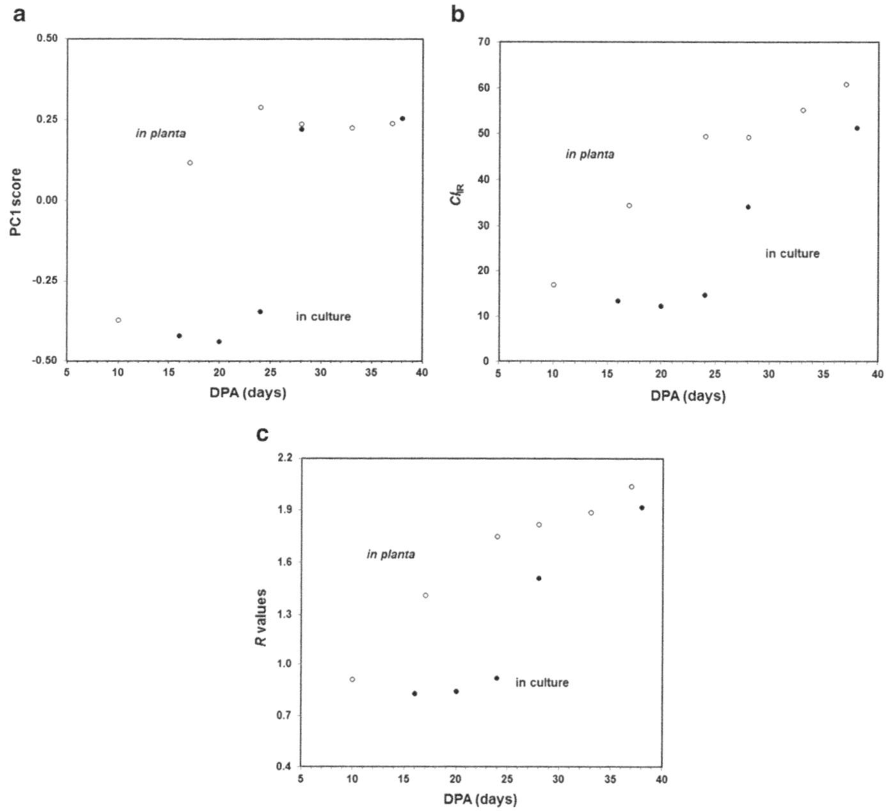

Fig. 4.2 (**a**) PC1 scores against DPA from normalized ATR FT-IR spectra of two fiber sets (Liu and Kim 2015). (**b**) CI_{IR} against DPAs from ATR FT-IR spectra of two fiber sets (Liu and Kim 2015). (**c**) *R* values against DPAs from ATR FT-IR spectra of two fiber sets (Liu and Kim 2015)

040 crystal reflections in natural fiber cellulose I, Hsieh's group calculated fiber crystallinity and crystallite size (Hsieh et al. 1997, 2000; Hu and Hsieh 2001). The results indicate that the cellulose I crystalline structure is clearly evident at the onset of the SCW formation (~20 DPA), and the 002 peak intensity increases greatly during cellulose biosynthesis, due to improved alignment of the glucosidic rings and improved order of atoms located within the glucosidic rings as the SCW thickens. For Maxxa fibers, fiber crystallinity increases steadily with fiber development from 38 to 57% between 24 and 60 DPA, while for the SJ-2 Acala fibers, it varies from 30 to 58% between 21 and 60 DPA (Hsieh et al. 1997). Compared to the crystallinity increasing gradually from 20 to 60 DPA for the Maxxa fibers, the crystallinity of the SJ-2 fibers reaches the maximum (~58%) at 34 DPA and remains virtually unchanged afterward. Although the apparent crystallite sizes normal to the 002 plane increase with fiber development for two types of fibers, the crystallite sizes between the two varieties are significantly different. Both fiber crystallinity and

crystallite size increase for developmental *G. hirsutum* (TM-1) and *G. barbadense* (Pima S7) cotton fibers (Hsieh et al. 2000). The crystallinity increases significantly between 20 and 35 DPA, and it does not change much beyond 35 DPA.

Other analytical tools have been explored in the structural and crystalline investigation of native cellulose I over the years. They are represented by solid-state ^{13}C nuclear magnetic resonance (NMR) spectroscopy (Hu and Hsieh 2001; Park et al. 2010), Raman spectroscopy (Schenzel et al. 2005), and IR spectroscopy (Abidi and Manike 2017; Kataoka and Kondo 1998; Liu et al. 2012; Nelson and O'Connor 1964a, b). By comparing the ^{13}C NMR spectra between hydrated and dried cotton fibers, Hu and Hsieh (2001) found that each resonance line of the hydrated cotton fiber is somewhat narrower than that of the dried cotton fiber without any change in chemical shifts and concluded that the noncrystalline chains in a hydrated state are relatively more ordered than those in a dried state. After resolving the resonance lines with a mixture of Gaussian and Cauchy functions and also calculating the integrated areas of the crystalline and noncrystalline components of the C4 resonance line, they reported 50% crystallinity for the dried fibers but 61% for the hydrated fibers. They pointed out that the degree of crystallinity estimated in this way is probably underestimated, because the crystallite surface also contributes to the C4 upfield wing. As confirmed in a separated study (Park et al. 2010), the C4 resonance region-based ^{13}C NMR spectral method produces lower crystallinity values than does the XRD method.

Schenzel et al. (2005) reported a FT-Raman spectroscopy for determining the degree of cellulose I crystallinity. They evaluated the *CI* of cellulose I samples by curve-fitting the crystallinity-dependent Raman CH_2 bending peaks at 1481 and 1462 cm^{-1}. The result shows that Raman crystallinity values for microcrystalline celluloses of different origins and varying crystallinity are reasonably correlated with those determined from NMR measurement.

Due to its rapid and nondestructive feature, IR spectroscopy has been implemented to assess cellulose crystallinity. In the 1960s, Nelson and O'Connor related the intensity ratios of the 1429 cm^{-1} band against the 893 cm^{-1} vibration (Nelson and O'Connor 1964a) or the intensity ratios of the 1372 cm^{-1} band against the 2900 cm^{-1} vibration (Nelson and O'Connor 1964b) to cellulose crystallinity and crystal lattice type. Recently, Liu et al. (2011, 2012) utilized IR bands at 708 and 730 cm^{-1}, indicative of respective I_β and I_α crystal forms, to estimate crystallinity index (CI_{IR}). In the study (Liu et al. 2012), they also proposed a simple ratio algorithm from major XRD peaks to determine the cellulose *CI* and then compared these CI_{XRD} ratios to the CI_{IR} from ATR FT-IR measurement. They concluded that a strong correlation between CI_{IR} and R3 (or CI_{XRD}) from a small set of samples could suggest the equivalence and effectiveness of the two separate measurements in fiber cellulose crystallinity determination (Fig. 4.3). They emphasized that this CI_{IR} procedure, which avoids the need to perform any pretreatment of the cotton fibers, has the advantage of using only small amounts of fiber (as little as 0.5 mg) and requires only a short time (less than 2 min) for sample loading, spectral acquisition, and subsequent result reporting.

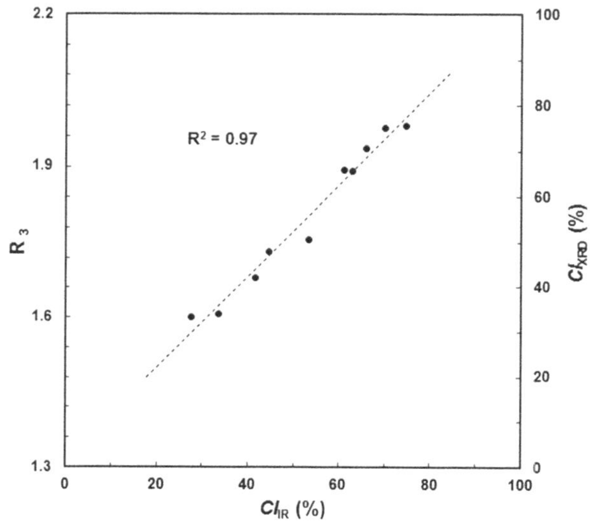

Fig. 4.3 CI_{IR} from FT-IR procedure vs. CI_{XRD} from XRD measurements (Liu et al. 2012)

Abidi and Manike (2017) investigated the structure and organization of cotton cellulose in two cultivars of developing *G. hirsutum* L. (TM-1 and TX55) cotton fibers using wide-angle X-ray diffraction (WAXD) and universal attenuated total reflectance Fourier transform infrared (UATR-FT-IR) spectroscopy. Their WAXD results show that a typical diffraction pattern is not apparent until 20 DPA and with fiber development, the intensity of the 200 peak increases which indicates improved alignment of the glycosidic rings. They observed very good linear relationship between the % crystallinity (determined by the Segal method) and the cellulose content (determined by the Anthrone method) and also between the % crystallinity and IR ratios and the integrated intensities of the 667 and 897 cm^{-1} vibrations (determined from UATR-FT-IR spectra).

4.4.5 Relating Cotton Fiber Strength with Crystallinity

Cotton fiber, as a raw and starting material, is primarily processed into yarns. Cotton fiber strength has been observed to have a high positive correlation with yarn strength and also to be the most important parameter influencing the yarn tenacity (Üreyen and Kadoglu 2006; Yang and Gordon 2017). The fundamental understanding of the relationship between cotton fiber strength and chemical structure (i.e., maturity and crystallinity) could be beneficial to cotton breeders for cotton crop enhancement and to textile processors for quality control. There are many studies on this topic (Benedict et al. 1994; Hsieh et al. 1997; Hu and Hsieh 2001; Islam et al. 2016; Liu et al. 2014). However, a direct comparison between these studies is difficult, because of differences in fiber strength measurement methods (Stelometer, HVI, or Instron tensile tester), crystallinity determination methods (XRD, ^{13}C NMR, or ATR FT-IR), and fiber sources (field grown or greenhouse grown). In the 1990s, Benedict et al. (1994) isolated crystalline microfibrillar fragments from different cotton fibers and reported a correlation (R) of 0.94 between the average length of the cellulose chains in the crystalline cellulose (measured from ^{13}C NMR spectra) and the HVI bundle fiber strength. Hu and Hsieh (2001) analyzed the XRD pattern to acquire the crystallite size and crystallinity information for the Acala variety of two upland cotton cultivars (SJ-2 and Maxxa) and then correlated the fiber breaking force and tenacity (or strength) from the Instron tensile tester with the crystallite size and crystallinity. Within each variety, single-fiber breaking forces were positively related to both the crystallite size and crystallinity, and increasing the breaking forces and tenacities appeared to be related more to crystallite size than to crystallinity. Meanwhile, the single-fiber breaking tenacities of the SJ-2 cotton fibers do not vary with fiber crystallinity, whereas those of Maxxa cotton fibers respond positively to an increase in crystallinity. They addressed that, besides the crystallinity and crystallite sizes, other structural parameters, such as fibril orientation and residual stress, may also play key roles in affecting the single-fiber strength of cotton fibers.

As a different strategy, Liu et al. (2014) utilized Stelometer instrument to determine the fiber bundle strength and then analyzed ATR FT-IR spectra on tiny Stelometer breakage specimens for fiber maturity and crystallinity. The Stelometer tested fiber bundles (2–5 mg) were too small for regular XRD pattern collection, which requires a relatively large amount of sample (~150 mg on an XRD aluminum holder of 25 mm diameter × 2 mm deep), but the specimens were sufficient for ATR FT-IR measurement. Compared to an obvious increase in fiber tenacity (or strength) with CI_{IR} for Pima fibers, there was an insignificant response between the two parameters for diverse upland fibers.

Islam et al. (2016) compared fiber properties of mature and developing fibers of *G. hirsutum* near-isogenic lines (NILs), MD52ne, MD90ne, and their F2 progenies, aiming to determine when variations of fiber properties occurred during cotton fiber development. They collected comprehensive fiber property and structural data using both novel and conventional methods that included Favimat, cross-sectional image analysis, ATR FT-IR, HVI, Stelometer, AFIS, and gravimetric fineness measurement. Their results show that the CI_{IR} is very low for the developing fibers between 10 and 15 DPA and then increases rapidly to 17 DPA before reaching its peak at 37 DPA. Developing MD52ne fibers at 20 DPA and older are significantly stronger than developing MD90ne fibers at the corresponding DPAs. When the developing fibers reached 20 DPA, the strength differences are clearly detected between MD52ne (21.7 g/tex) and MD90ne (17.5 g/tex). Overall, strengths of developmental fibers with 20, 24, 28, 33, 37, and 44 DPA increase with crystallinity for 2 NILs.

4.4.6 Chemical Imaging Characterization of Cotton Fiber Cellulose

Chemical imaging technique, a combination of conventional one-dimensional spectral information and spatial information, has been utilized to examine cotton fibers at single-fiber level (Himmelsbach et al. 2003; Santiago et al. 2016, 2017). The results show that the spectral changes observed in FT-IR microspectroscopy/imaging measurement generally coincide with those observed with macroscopic sampling techniques, such as ATR FT-IR. PCA scatter plot of the microspectroscopy/imaging data allows for the facile distinction of cotton fibers at different developmental time points. One significant advantage of the microspectroscopy method lies in the calculation of the chemical distribution maps that can visually depict general spectral changes (Santiago et al. 2017).

4.4.7 Estimating Mesoscale Packing of Cellulose Microfibrils

An innovate technique known as vibrational sum-frequency generation (SFG) spectroscopy was explored to study the mesoscale packing and crystal structure of cellulose microfibrils as well as temporal changes in cell wall composition and hydration in two developmental cotton fiber species (*G. hirsutum* cv Deltapine 90 vs. *G. barbadense* cv Phytogen 800) (Lee et al. 2015). SFG analysis of fibers from unopened bolls showed subtle differences between two species in the mesoscale ordering of cellulose microfibrils in the maturing SCW areas. Among the mature fibers dried on the plant after the boll opened naturally, the difference in SFG spectra between two species was insignificant. PCA characterization of ATR FT-IR and FT-Raman spectra reveals that fibers of both species undergo abrupt changes in the cellulose and matrix polymer contents during the transition to SCW synthesis, and XRD result suggests that cellulose crystal size and crystallinity increase similarly over the growth time in both species.

Contrary to the earlier report (Wang et al. 2014) that the growth in immature fiber (*im*) mutant fibers might be delayed, a comparative analyses, by applying HVI, AFIS, gravimetric fiber fineness, cellulose assay, cross-sectional and polarized microscopy, Cottonscope, vibrational spectroscopy (Raman, FT-IR, and SFG), and XRD measurement, show that *im* and TM-1 fibers exhibit the similar developmental process of SCW thickening (Kim et al. 2017). Although there is no significant difference detected in cellulose content, crystallinity index, crystal size, and ribbon width between two types of fibers, there are significant differences observed in linear density, cross-sectional micrographs, and the lateral order between cellulose microfibrils (CMFs). The linear density was lower, and the CMFs were less organized in *im* fibers than the TM-1 fibers.

4.5 Determination of Cotton Fiber Noncellulosic Components and Their Implications

Noncellulosic components in cotton fibers have also been investigated considerably, although not in the same level of attention as the cellulose component.

4.5.1 *Wax*

Cotton wax is one of the noncellulosic components inside the cuticle and PCW area on the outer surface of the cotton fiber. It is highly complex in chemical constitutes and is composed of a mixture of organic substances such as waxy alkanes, fatty acids, fatty alcohols, and plant steroids (Church and Woodhead 2006, 2017). The amount of cotton wax was determined by the Conrad method that involves multiple extraction and separation steps (Conrad 1944). These extracts were further analyzed by various chemical instruments, including thin-layer chromatography (TLC), liquid chromatography, gas chromatography/mass spectrometry (GC/MS), and IR spectroscopy, for the identification and quantitation of specific and individual wax species.

Surface-sensitive techniques, such as ATR FT-IR and photoacoustic FT-IR (PAS FT-IR) spectroscopy, were used to study cotton wax (Church and Woodhead 2017). For raw cotton fiber, the appearance of very weak but sharp IR band near 2900 cm^{-1} due to the C–H stretching mode is an indicative of the wax presence. PCA processing of PAS FT-IR spectra revealed that scoured and unscoured cotton fibers were easily identified from each other and alkane content was responsible for the separation (Church and Woodhead 2006). These spectroscopic results are in good agreement with those obtained by TLC analysis.

ATR mapping utilizing a focal plane array allows for the visualization of the entire surface at a single cotton fiber level and their convoluted surfaces at a spatial resolution of 1.56 μm (Church and Woodhead 2017). By analyzing the unique bands at 2915 and 2850 cm^{-1} assigned to the asymmetric and symmetric CH_2 stretching mode of the aliphatic hydrocarbon chains in the cotton wax, the result suggests the potential of ATR mapping in providing micron-level information about the chemical composition and distribution of cotton wax at fiber surface (Church and Woodhead 2017).

Cotton wax on the fiber surface was also characterized by surface energy analyzer-inverse gas chromatography (SEA-IGC) (Rjiba et al. 2007). It clearly showed that the presence of wax on cotton fibers affects the dispersive surface energy and the acid-base character of the fibers.

Cotton wax is most responsible for the hydrophobicity and low water wettability of raw cotton fiber and also for the protection of cotton fibers from microbial degradation of the underlying carbohydrates. It plays an important role as a lubricant in reducing fiber damage during the proper spinning of cotton fibers into yarn (Gamble 2004), but its content does not show a significant correlation with fiber properties, fiber friction, and fiber breakage in textile processing (Cui et al. 2002; Gamble 2004). However, increases in yarn strength were found to be correlated with elevation of total fiber alcohol surface extractables, wax, and potassium content (Brushwood 2005). Single-skein yarn tensile tests reveal a significant increase in yarn tenacity (or strength) following the yarn dewaxing process by Soxhlet extraction with ethanol, and yarn strength was explained primarily by fiber strength and fiber fineness, not wax content, in regression equations for yarn strength (Price et al. 2002).

4.5.2 Sugars

Plant sugars and insect sugars are two sources of soluble sugars in cotton fibers. Plant sugars (or metabolic sugars) reside within the lumen and the outer fiber surface resulted from the growth process, and they are represented by dominant monosaccharides, glucose, and fructose. Insect sugars, commonly known as honeydew, mainly come from whiteflies and aphids. High levels of either plant or inset sugars in cotton fibers cause fiber stickiness (Gamble 2003a; Hequet and Abidi 2006). The sugars in sticky cotton fibers build up very quickly on rollers and saws in cotton gins and various components in the spinning mills, leading to the contamination of processing equipments. Presence of physiological sugars that cause the stickiness in developing fibers also affects the accurate fineness value measurement by AFIS protocol (Kim et al. 2014).

In the laboratory, multiple chemical and physical methods have been developed to measure cotton stickiness in meeting the industry need (Abidi and Hequet 2007; Barton et al. 2005; Gamble 2003a; Hequet and Abidi 2006; Peralta et al. 2016, 2017). As one approach, Barton et al. (2005) established FT-NIR models to sense cotton stickiness and to measure the sugars present on the basis of reference values from HPLC, thermodetector, and Minicard measurement. They reported a successful model to determine the Minicard value and concluded that FT-NIR technique can be used to detect stickiness in raw cotton fibers. Meanwhile, they observed a poor performance of the sugar models, indicating larger oligomers and/or proteins other than simple sugars are important to model stickiness. In a preliminary study, Abidi and Hequet (2007) demonstrated the usefulness of the UATR-FT-IR in detecting the presence of trehalulose (the dominant sugar in the whitefly honeydew residue). They observed that the integrated intensities of the peaks located at 3280, 1622, and 1018 cm^{-1} show high correlation with the trehalulose content on the lint, and also it is possible to discriminate between contaminated and non-contaminated cotton fibers through PCA interpretation of the UATR-FT-IR spectra.

Recently, Peralta et al. (2016, 2017) reported the use of ion chromatography (IC) to identify and quantify 9 sugars present in the water extracts of 25 cotton samples in order to create a sugar profile for each sample, since specific levels and ratios of the carbohydrates melezitose and trehalulose deposited on the surface of cotton fibers are indicators of whitefly or aphid contamination. They compared the sugar profiles for each sample to its respective Minicard rating of none, light, moderate, or heavy in order to distinguish a meaningful relationship between the IC data and the Minicard rating. Trehalulose and melezitose in the water extracts have the highest positive correlation to the Minicard rating, whereas glucose and fructose have high negative correlations. There are relatively little changes in the sugar distribution ratios between water-extracted sugars and the sugars contained in the sticky spots.

Sugars, together with electrolytes and pectin, show significant correlations with single-fiber friction, and the removal of sugars and electrolytes by water washing the fibers does not produce a significant change in frictional behavior (Gamble 2004). Glucose content was reported to decrease substantially in aged fibers, and its reduction is associated with a concomitant decrease in pH as a result of the production of acidic metabolic by-products due to anaerobic microbial degradation of glucose (Gamble 2003b).

4.5.3 Pectins and Hemicelluloses

Underlying the waxy cuticle is the PCW, which is considered to have two distinctive layers. The outer layer consists of pectin derivatives (simply as pectins) mainly in the form of pectic acid and rhamnose, while the inner layer is comprised of hemicelluloses, typically in the form of xyloglucan, and cellulose. Like waxes, pectins play a role for cotton fiber hydrophobicity and can be removed by the scouring process or pectinase enzymes. Pectin content was determined by enzymatic degradation of the cotton fiber and the subsequent analysis of galacturonic acid by high-performance anion-exchange chromatography (HPAEC) method (Gamble 2004). Of the four chemical components (pectin, wax, soluble salts, and glucose) examined in this study, increasing pectin content appears to be the most strongly associated with decreasing inter-fiber friction.

4.5.4 Proteins

The nitrogen-containing compounds, expressed as percent proteins, are presented in the lumen and PCW area. Gamble (2008) suggested that a chemical condensation between a reducing sugar and an amino acid, or the Maillard reaction, induces an increase in cotton fiber yellowness (+b) that impacts commercial cotton fiber market value. In his work, extracted sugars (glucose and fructose) and pH were measured in oven-treated cotton fibers in order to develop a model capable of predicting +b changes at any given temperature and time. Due to their low amounts in mature cotton fibers, there is no literature available on the effect of proteins/amino acids on fiber friction and fiber/yarn physical properties.

4.5.5 *Inorganics*

Inorganic substances exist as metabolic residues in the form of simple organic salts or inorganic anions. They are essential for the growth of cotton plant as normal nutrients. Besides the metals absorbed by cotton fibers, soil and wind-borne dust could be transferred into the surface of cotton fibers, especially during the mechanic harvesting process. In general, the levels of these matters are very low, but they may contribute to problems in yarn manufacturing, bleaching, and dyeing. Brushwood (2005) correlated noncellulosic materials on raw cotton fibers with cotton fiber spinning performance. Of his findings include the increase in yarn strength with increasing potassium content and also the increase in fiber yellowness and grayness influenced by increasing potassium, iron, or magnesium content. Gamble (2009) measured the contents of eight water-soluble metal cations (calcium, potassium, sodium, magnesium, iron, nickel, copper, and zinc) in water extraction, scouring treatment, and microwave digestion solutions of cotton fibers by inductively coupled plasma-optical emission spectroscopy (ICP-OES). The results show that the K^+ (potassium ion) content is primarily determined by the amount of rainfall prior to harvest and by fiber maturity, (1) the Ca^{2+} (calcium ion) amount in the scour solution is directly and negatively correlated with fiber maturity, and the Ca^{2+} and Mg^{2+} (magnesium ion) amount remains present in the post-scoured fiber, with an indication that lower maturity cotton fibers retain higher levels of the Ca^{2+} amount than do more mature cotton fibers.

Brown cotton (*G. hirsutum* L.) fibers (SA-1 and MC-BL) were reported to be inferior to white cotton fiber Sure-Grow 747 (SG747) in fiber quality (i.e., shorter length and fewer twists) and chemical structure (crystallinity), but they exhibited superior thermal properties determined through the thermogravimetry (TG), differential thermogravimetry (DTG), and microscale combustion calorimetry (MCC) analyses (Nam et al. 2016). The result shows that rich natural inorganic components and condensed tannins present in brown cotton fibers are responsible for their thermal stability. Among inorganic elements, the logarithmic concentration of sodium ion (Na^+) was observed to have a significant correlation with heat release capacity. The condensed tannins induce strong binding with Na^+ and exhibit the thermal-resistant characteristics.

4.6 Conclusion

Characteristics of cotton fiber composition and structure are ongoing challenge mostly due to its complexity. Understanding the fiber composition and structure is very important in order to enhance the cotton fiber production, utilization, and processing. The reality of limited cellulose solubility in most solvents makes it a hindrance to characterize fiber cellulose rapidly and accurately. As a complementary approach to traditional chemical instrumental methods, FT-IR spectroscopy, a rapid and nondestructive technique, has been proved to be a powerful tool to study cotton fiber cellulose composition and structure for cotton fiber physiology and breeding applications. The accumulated knowledge will enhance cotton cultivar selection in physiology and breeding applications, ultimately aiding to understand the fiber structure and end-use quality in combination with diverse research and comprehensive fiber quality test tools.

References

Abidi N, Cabrales L, Haigler CH (2014) Changes in the cell wall and cellulose content of developing cotton fibers investigated by FTIR spectroscopy. Carbohydr Polym 100:9–16

Abidi N, Cabrales L, Hequet E (2010) Fourier transform infrared spectroscopic approach to the study of the secondary cell wall development in cotton fiber. Cellulose 17:309–320

Abidi N, Hequet E (2007) Fourier transform infrared analysis of cotton contamination. Text Res J 77:77–84

Abidi N, Hequet E, Cabrales L (2010) Changes in sugar composition and cellulose content during the secondary cell wall biogenesis in cotton fibers. Cellulose 17:153–160

Abidi N, Hequet E, Cabrales L, Gannaway J, Wilkins T, Wells LW (2008) Evaluating cell wall structure and composition of developing cotton fibers using Fourier transform infrared spectroscopy and thermogravimetric analysis. J Appl Polym Sci 107:476–486

Abidi N, Manike M (2017) X-ray diffraction and FTIR investigations of cellulose deposition during cotton fiber development. Text Res J 88:719–730

Barton FE II, Bargeron JD III, Gamble GR, McAlister DL, Hequet E (2005) Analysis of sticky cotton by near-infrared spectroscopy. Appl Spectrosc 59:1388–1392

Benedict C, Kohel R, Jividen G (1994) Crystalline cellulose and cotton fiber strength. Crop Sci 34:147–151

Boylston EK, Hebert JJ (1995) The primary wall of cotton fibers. Text Res J 65:429–431

Bradow JM, Davidonis GH (2000) Quantitation of fiber quality and the cotton production–processing interface: a physiologist's perspective. J Cotton Sci 4:34–64

Brushwood DE (2005) Predicting yarn processing performance from the noncellulosic content of raw cottons. Text Res J 75:1–5

Cho Y, Han Y, Lambert W, Bragg C (1996) Characterizing convolutions in cotton fiber and their effects on fiber strength. Trans ASAE 40:479–483

Church JS, Woodhead AL (2006) Spectroscopic assessment of Australian cotton waxes. Appl Spectrosc 60:1334–1340

Church JS, Woodhead AL (2017) Cotton fiber wax and surface properties. In: Gordon S, Abidi N (eds) Cotton fibers: characteristics, uses and performance. Nova Science, New York, pp 21–41

Conrad CM (1944) Determination of wax in cotton fiber a new alcohol extraction method. Ind Eng Chem Anal Ed 16:745–748

Cui XL, Price JB, Calamari TA, Hemstreet JM, Meredith WR (2002) Cotton wax and its relationship with fiber and yarn properties: part I: wax content and fiber properties. Text Res J 72: 399–404

Fang DD, Percy RG (2015) Cotton, Agronomy monograph 57, 2nd edn. American Society of Agronomy, Crop Science Society of America, and Soil Science Society of America, Madison

Foulk JA, Mcalister DD III (2002) Single cotton fiber properties of low, ideal, and high micronaire values. Text Res J 72:885–891

Frydrych I, Thibodeaux D (2010) Fiber quality evaluation – current and future trends/intrinsic value of fiber quality in cotton. In: Wakelyn PJ, Chaudhry MR (eds) Cotton: technology for the 21st century. International Cotton Advisory Committee, Washington, pp 251–295

Gamble GR (2003a) Evaluation of cotton stickiness via the thermochemical production of volatile compounds. J Cotton Sci 7:45–50
Gamble GR (2003b) Variation in surface chemical constituents of cotton (*Gossypium hirsutum*) fiber as a function of maturity. J Agric Food Chem 51:7995–7998
Gamble GR (2004) Implications of surface chemistry on cotton fiber processing. J Cotton Sci 8:198–204
Gamble GR (2008) Method for the prediction of the rate of +b color change in upland cotton (*Gossypium hirsutum* L.) as a function of storage temperatures. J Cotton Sci 12:171–177
Gamble GR (2009) Regional, varietal, and crop year variations of metal contents associated with the separate structural components of upland cotton (*Gossypium hirsutum*) fiber. J Cotton Sci 13:221–226
Gordon S, Abidi N (2017) Cotton fibers: characteristics, uses and performance. Nova Science, New York
Gordon S, Hsieh Y-L (2007) Cotton: science and technology. Woodhead Publishing, Cambridge
Hartzell-Lawson MM, Hsieh Y-L (2000) Characterizing the noncellulosics in developing cotton fibers. Text Res J 70:810–819
Hearle JWS (2007) Physical structure and properties of cotton. In: Gordon S, Hsieh Y-L (eds) Cotton: science and technology. Woodhead Publishing, Cambridge, pp 35–67
Hearle JWS, Sparrow JT (1971) The fractography of cotton fibers. Text Res J 41:736–749
Hequet EF, Abidi N (2006) Sticky cotton: measurements and fiber processing. Texas Tech University Press, Lubbock, pp 14–29
Himmelsbach DS, Akin DE, Kim J, Hardin IR (2003) Chemical structural investigation of the cotton fiber base and associated seed coat: Fourier-transform infrared mapping and. Text Res J 73:281–288
Hindeleh AM (1980) Crystallinity, crystallite size, and physical properties of native Egyptian cotton. Text Res J 50:667–674
Hsieh Y-L (2007) Chemical structure and properties of cotton. In: Gordon S, Hsieh Y-L (eds) Cotton: science and technology. Woodhead Publishing, Cambridge, pp 3–34
Hsieh Y-L, Hu X-P, Nguyen A (1997) Strength and crystalline structure of developing Acala cotton. Text Res J 67:529–536
Hsieh Y-L, Hu X-P, Wang A (2000) Single fiber strength variations of developing cotton fibers. Text Res J 70:682–690
Hu X-P, Hsieh Y-L (2001) Effects of dehydration on the crystalline structure and strength of developing cotton fibers. Text Res J 71:231–239
Huwyler HR, Franz G, Meier H (1979) Changes in the composition of cotton fiber cell walls during development. Planta 146:635–642
Islam MS, Fang DD, Thyssen GN, Delhom CD, Liu Y, Kim HJ (2016) Comparative fiber property and transcriptome analyses reveal key genes potentially related to high fiber strength in cotton (*Gossypium hirsutum* L.) line MD52ne. BMC Plant Biol 16:36
Kataoka Y, Kondo T (1998) FT-IR microscopic analysis of changing cellulose crystalline structure during wood cell wall formation. Macromolecules 31:760–764
Kelly C, Hequet E, Dever J (2012) Interpretation of AFIS and HVI fiber property measurements in breeding for cotton fiber quality improvement. J Cotton Sci 16:1–16
Kim HJ (2015) Fiber biology. In: Fang DD, Percy RG (eds) Cotton, Agronomy monograph 57, 2nd edn. American Society of Agronomy, Crop Science Society of America, and Soil Science Society of America, Madison, pp 97–127
Kim HJ, Lee CM, Dazen K, Delhom CD, Liu Y, Rodgers JM, French AD, Kim SH (2017) Comparative physical and chemical analyses of cotton fibers from two near isogenic upland lines differing in fiber wall thickness. Cellulose 24:2385–2401
Kim HJ, Rodgers J, Delhom C, Cui X (2014) Comparisons of methods measuring fiber maturity and fineness of upland cotton fibers containing different degree of fiber cell wall development. Text Res J 84:1622–1633

Lee CM, Kafle K, Belias DW, Park YB, Glick RE, Haigler CH, Kim SH (2015) Comprehensive analysis of cellulose content, crystallinity, and lateral packing in *Gossypium* hirsutum and *Gossypium* barbadense cotton fibers using sum frequency generation, infrared and Raman spectroscopy, and X-ray diffraction. Cellulose 22:971–989

Lee JJ, Woodward AW, Chen ZJ (2007) Gene expression changes and early events in cotton fibre development. Ann Bot 100:1391–1401

Li C, Guo W, Zhang T (2009) Fiber initiation development in upland cotton (*Gossypium hirsutum* L.) cultivars varying in lint percentage. Euphytica 165:223–230

Liu Y (2015) Rapid and routine assessment of cotton fiber cellulose maturity: current and future trends. In: Mondel MIH (ed) Cellulose and cellulose derivatives: synthesis, modification and applications. Nova Science, New York, pp 17–25

Liu Y, Kim HJ (2015) Use of attenuated total reflection Fourier transform infrared (ATR FT-IR) spectroscopy in direct, non-destructive, and rapid assessment of developmental cotton fibers grown in planta and in culture. Appl Spectrosc 69:1004–1010

Liu Y, Kim HJ (2017) Fourier transform infrared spectroscopy (FT-IR) and simple algorithm analysis for rapid and non-destructive assessment of developmental cotton fibers. Sensors 17:1469

Liu Y, Thibodeaux D, Gamble G (2011) Development of Fourier transform infrared spectroscopy in direct, non-destructive, and rapid determination of cotton fiber maturity. Text Res J 81:1559–1567

Liu Y, Thibodeaux D, Gamble G, Bauer P, VanDerveer D (2012) Comparative investigation of Fourier transform infrared (FT-IR) spectroscopy and X-ray diffraction (XRD) in the determination of cotton fiber crystallinity. Appl Spectrosc 66:983–986

Liu Y, Thibodeaux D, Gamble G, Rodgers J (2014) Preliminary study of relating cotton fiber tenacity and elongation with crystallinity. Text Res J 84:1829–1839

Lord E (1956) Air flow through plugs of textile fibers part II. The micronaire test for cotton. J Text Inst Trans 47:T16–T47

Meinert MC, Delmer DP (1977) Changes in biochemical composition of the cell wall of the cotton fiber during development. Plant Physiol 59:1088–1097

Nam S, Kim HJ, Condon BD, Hinchliffe DJ, Chang S, Mccarty JC Jr, Madison CA (2016) High resistance to thermal decomposition in brown cotton is linked to tannins and sodium content. Cellulose 23:1137–1152

Nelson ML, O'Connor RT (1964a) Relation of certain infrared bands to cellulose crystallinity and crystal lattice type. Part I. Spectra of lattice types I, II, and III and of amorphous cellulose. J Appl Polym Sci 8:1311–1324

Nelson ML, O'Connor RT (1964b) Relation of certain infrared bands to cellulose crystallinity and crystal lattice type. Part II. A new infrared ratio for estimation of crystallinity in cellulose I and II. J Appl Polym Sci 8:1325–1341

Park S, Baker JO, Himmel ME, Parilla PA, Johnson DK (2010) Cellulose crystallinity index: measurement techniques and their impact on interpreting cellulase performance. Biotechnol Biofuels 3(1):10

Paudel D, Hequet E, Abidi N (2013) Evaluation of cotton fiber maturity measurements. Ind Crop Prod 45:435–441

Peralta DV, Fortier CA, Thibodeaux D, Delhom CD, Rodgers JE III (2016) Separation and quantitation of plant and insect carbohydrate isomers found on the surface of cotton. AATCC J Res 3:13–23

Peralta DV, Fortier CA, Zumba J, Thibodeaux D, Delhom CD, Rodgers JE III (2017) Comparisons of minicard ratings to ion chromatography sugar profiles in cotton fiber water extract and minicard sticky spot material. Text Res J 87:747–758

Price JB, Cui XL, Calamari TA, Meredith WR (2002) Cotton wax and its relationship with fiber and yarn properties, part II: wax content and yarn properties. Text Res J 72:631–637

Rjiba N, Nardin M, Drean JY, Frydrych RA (2007) A study of the surface properties of cotton fibers by inverse gas chromatography. J Colloid Interface Sci 314:373–380

Romano GB, Taliercio EW, Turley RB, Scheffler JA (2011) Fiber initiation in 18 cultivars and experimental lines of three *Gossypium* species. J Cotton Sci 15:61–72

Santiago CM, Fortier CA, Hinchliffe DJ, Rodgers JE III (2017) Chemical imaging of secondary cell wall development in cotton fibers using a mid-infrared focal-plane array detector. Text Res J 87:1040–1051

Santiago CM, Hinchliffe DJ (2015) FT-IR examination of the development of secondary cell wall in cotton fibers. Fibers 3:30–40

Santiago CM, Hinchliffe DJ, Montalvo JG Jr, Von Hoven TM, Rodgers JE III, Thyssen GN, Zeng L, Madison CA (2016) Infrared imaging of cotton fiber bundles using a focal plane array detector and a single reflectance accessory. Fibers 4:27

Schenzel K, Fischer S, Brendler E (2005) New method for determining the degree of cellulose I crystallinity by means of FT Raman spectroscopy. Cellulose 12:223–231

Segal L, Creely JJ, Martin AE, Conrad CM (1962) An empirical method for estimating the degree of crystallinity of native cellulose using the X-ray diffractometer. Text Res J 29:786–794

Singh B, Avci U, Eichler Inwood SE, Grimson MJ, Landgraf J, Mohnen D, Sorenson I, Wilkerson CG, Willats WGT, Haigler CH (2009) A specialized outer layer of the primary cell wall joins elongating cotton fibers into tissue-like bundles. Plant Physiol 150:684–699

Stiff MR, Haigler CH (2012) Recent advances in cotton fiber development. In: Oosterhuis DM, Cothren JT (eds) Flowering and fruiting in cotton. The Cotton Foundation, Cordova, pp 163–192

Thibodeaux DP, Evans JP (1986) Cotton fiber maturity by image analysis. Text Res J 56:130–139

Timpa JD, Triplett BA (1993) Analysis of cell-wall polymers during cotton fiber development. Planta 189:101–108

Updegraff DM (1969) Semimicro determination of cellulose in biological materials. Anal Biochem 32:420–424

Üreyen ME, Kadoglu H (2006) Regressional estimation of ring cotton yarn properties from HVI fiber properties. Text Res J 76:360–366

Viles FJ, Silverman L (1949) Determination of starch and cellulose with anthrone. Anal Chem 21:950–953

Wakelyn PJ, Bertoniere NR, French AD, Thibodeaux DP, Triplett BA, Rousselle AA, Goynes WR Jr, Edwards JV, Hunter L, Mcalister DD, Gamble GR (2007) Cotton fiber chemistry and technology. CRC Press, Boca Raton

Wakelyn PJ, Bertoniere NR, French AD, Thibodeaux DP, Triplett BA, Rousselle AA, Goynes WR Jr, Edwards JV, Hunter L, Mcalister DD, Gamble GR (2007a) Cotton fiber chemistry and technology. CRC Press, Boca Raton, pp 11–14

Wakelyn PJ, Bertoniere NR, French AD, Thibodeaux DP, Triplett BA, Rousselle AA, Goynes WR Jr, Edwards JV, Hunter L, Mcalister DD, Gamble GR (2007b) Cotton fiber chemistry and technology. CRC Press, Boca Raton, pp 15–20

Wakelyn PJ, Bertoniere NR, French AD, Thibodeaux DP, Triplett BA, Rousselle AA, Goynes WR Jr, Edwards JV, Hunter L, Mcalister DD, Gamble GR (2007c) Cotton fiber chemistry and technology. CRC Press, Boca Raton, pp 23–68

Wang C, Lv Y, Xu W, Zhang T, Guo W (2014) Aberrant phenotype and transcriptome expression during fiber cell wall thickening caused by the mutation of the *im* gene in immature fiber (*im*) mutant in *Gossypium hirsutum* L. BMC Genomics 15:94

Yang S, Gordon S (2017) Accurate prediction of cotton ring-spun yarn quality from high-volume instrument and mill processing data. Text Res J 87:1025–1039

Chapter 5
Chemical Properties of Cotton Fiber and Chemical Modification

Noureddine Abidi

5.1 Chemical Properties

Cotton fibers are composed of about 95% of cellulose after ginning and mechanical cleaning. Cellulose is a polysaccharide composed of β-D glucopyranose units covalently linked through 1,4-glycosidic bonds. The number of repeating glucose units, called degree of polymerization, varies with the origin (cultivar) of the raw cotton fibers as well as with the treatment (Klemm et al. 2005; Park et al. 2010).

Cellobiose is the union of two glucose molecules through glycosidic bond (Cocinero et al. 2009). Each successive glucose unit in cellulose macromolecule is rotated 180° around the molecular axis. Each glucopyranose unit of cellulose contains three hydroxyl groups. The hydroxyl groups present on C-6 position are called primary hydroxyl groups, and those on C-2 and C-3 are called secondary hydroxyl groups (Fig. 5.1). The primary group is more easily neutralized and readily available and prominent in chemical reactions of cellulose. Hydroxyl groups are readily involved in intra- and intermolecular hydrogen bonds and lead to ordered crystalline arrangements. Cellulose molecules in cotton are organized into parallel arrangements called crystallites and, subsequently, into larger aggregates called fibrils (Maxwell et al. 2003). Thirty-six cellulosic chains are assembled to compose an elementary fibril (diameter ~3.5 nm). These elementary fibrils, which are taken as the basic crystalline unit of cotton cellulose, are assembled into microfibrils with a diameter varying from 10 to 30 nm (Lee et al. 2000; Peterlin and Ingram 1970). Between the ordered crystalline regions, cellulosic chains are randomly distributed, and these regions are called amorphous regions. While crystalline regions are responsible for the required strength of the fiber, amorphous regions are essential

N. Abidi (✉)
Fiber and Biopolymer Research Institute, Department of Plant and Soil Science, Texas Tech University, Lubbock, TX, USA
e-mail: noureddine.abidi@ttu.edu

D. D. Fang (ed.), *Cotton Fiber: Physics, Chemistry and Biology*,
https://doi.org/10.1007/978-3-030-00871-0_5

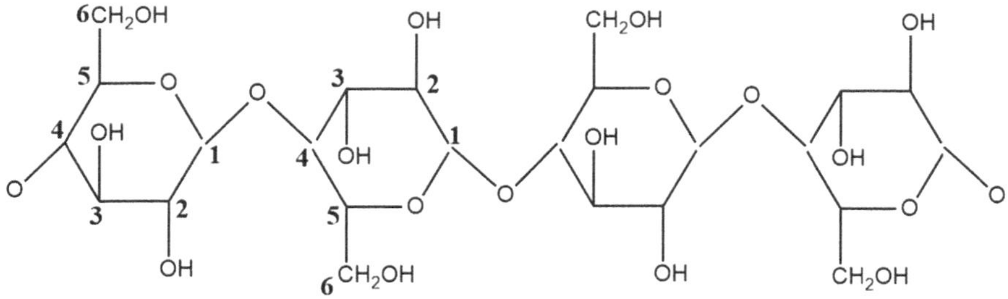

Fig. 5.1 Chemical structure of cellulose

for elasticity, flexibility, and the ability of the fibers to absorb water, dyes, and chemical finishes (Rivlin 1992). Approximately 70–80% of the cellulose in cotton is in crystalline form I_{β}.

5.2 Cotton Fabric Preparation

In general, cotton fibers are processed (opening, carding, drawing, spinning, and weaving/knitting) in their raw state. The presence of the primary cell wall containing wax and pectins provides the necessary lubrication and prevents static electricity buildup during processing. If waving is desired, yarn is coated with starch or polyvinyl alcohol (this process is termed sizing) to give it stiffness and allow easy waving. The sizing material (starch or polyvinyl alcohol) must be removed when the woven fabric is produced. The process of desizing is then performed, which consists of boiling cotton fabric in water bath containing nonionic wetting agent, hydrogen peroxide, and acetic acid. The desizing process is normally performed at 194 °F for 30 min followed by neutralization with acetic acid. If the size is not removed or partly removed, the remaining starch or polyvinyl alcohol could interfere with the application of the dye and chemical finishing, which may affect the properties of the end product.

Desizing starch can be also achieved using amylase enzymes (Spicka and Tavcer 2013; Eren et al. 2009). Four groups of starch-degrading enzymes exist (Eren et al. 2009): endoamylases, exoamylases, debranching enzymes, and transferases. Typical enzymatic desizing bath contains 10 ml/L of α-amylases, 10 mL/L of amyloglucosidases, and 10 mL/L of pullulanases (Spicka and Tavcer 2013).

To remove noncellulosic compounds present in/on the cotton fibers and expose hydroxyl groups of cellulose to further chemical reaction, a scouring process is performed. The scouring process is normally performed in a solution containing wetting agent (example of wetting agent is Triton X-100), caustic soda, and acetic acid. This treatment solubilizes impurities present on/in cotton fibers. The scouring process, also referred to as alkali boil-off, is typically performed with a strong alkali solution (5–10 g/L sodium hydroxide) close to or above the boiling temperature of water for 1–2 h (Rivlin 1992). The fabric is subjected to hot rinse to remove emulsified impurities followed by neutralization with acetic acid (Rivlin 1992).

To further remove and decompose colored impurities, a bleaching process is performed. This process is performed using hydrogen peroxide and/or sodium chloride. A typical bleaching process is normally performed near the boil or at temperature above the boil using a bath composed of wetting agent, caustic soda (0.5–1% on the weight of the fabric), sodium silicate (2–4% on the weight of the fabric), hydrogen peroxide 35% (4–8% on the weight of the fabric), and acetic acid. A final step of fabric neutralization is done with acetic acid (Rivlin 1992).

An optional treatment is sometimes performed which consists of treating cotton yarn or fabric with concentrated solution of sodium hydroxide (NaOH). This process is referred to as mercerization. The typical mercerization process requires immersing cotton yarn or fabric in caustic soda (20–25%) for brief period of time (less than 5 min) usually under tension followed by rinsing, boiling, and neutralization. The mercerization process has many beneficial effects such as increased tensile strength, increased softness, improved affinity to dyes, improved dyeability of immature fibers, increased luster, and higher water sorption. Changes in the microstructure of fibers occur leading to conversion of the crystal structure from cellulose I to cellulose II and reduction in crystallite size (Rivlin 1992).

5.3 Water/Oil Repellency and Self-Cleaning

The presence of hydroxyl groups on cellulose macromolecules imparts hydrophilicity to cotton fabrics. This property can be changed by altering the surface tension of the fabric to impart water repellency. Changes in the surface tension prevent water from diffusing inside the fibers. The ability of the textile substrate to repel a liquid is assessed by performing water contact angle measurements. In general, a textile fabric is classified as hydrophilic, hydrophobic, or super hydrophobic depending on the values of the contact angles. When the static water contact angle is $<90°$, the textile surface is hydrophilic. When the contact angle is $>90°$ but $<150°$, the surface is classified as being hydrophobic. When the contact angle is $>150°$, the textile surface is said to be super hydrophobic. Reducing the interaction between a drop of a liquid and a textile surface means that the water contact angle is considerably increased. This allows the liquid droplet to easily roll off the surface, which can easily carry with it the dirt on surface. This will impart "self-cleaning properties" to the surface. Functionalization of textile substrates to impart water repellency is achieved by using hydrophobic compounds such as silicones (Liu et al. 2009b; Hsieh et al. 2010) and fluorocarbons (Di Mundo et al. 2009; Favia et al. 2003; Kulinich and Farzaneh 2005; Liu et al. 2009a). However, the drawbacks of these approaches are disposal of solvents, inhomogeneous distribution of the reactants on the surface, control of the reaction, and generation of toxic effluents. To overcome these drawbacks, the focus has been on using a "dry process" such as plasma technology and molecular vapor deposition.

5.3.1 Plasma Modification

The plasma is composed of radicals, metastable molecules, photons, and charged particles such as ions and electrons (Abidi and Hequet 2004). Plasma is generally generated by exposing a gas (such as Ar, N_2, O_2, etc.) to radiofrequency or microwave electrical discharges. Because plasma density is proportional to the square root of the frequency, the degree of ionization increases with increasing frequency. Microwave plasma produces higher density of reactive particles compared to radiofrequency. Plasma technology is attractive for surface modification of textile substrates. Its advantage is that no toxic radicals are used as initiators or cross-linkers, and the amount of toxic effluents is minimized. In addition, the temperature of the plasma-treated substrate does not increase, which makes the plasma process a cold process. No thermal degradation of the textile substrate occurs. Furthermore, the effect of the plasma does not penetrate more than 100 nm from the surface. Therefore, the cotton fabric retains most of the desirable physical properties (Abidi and Hequet 2004; Bhat and Benjamin 1999).

Abidi and Hequet reported on the impact of the microwave plasma on the physical properties of cotton fabric (Abidi and Hequet 2005). In their experiments, the authors exposed cotton fabric to Ar-plasma for a period of time varying between 60 and 400 s and at varying microwave power between 100 and 500 W. the results showed that exposing cotton fabric to Ar-plasma for 240 s at 500 W resulted in only 0.4% weight loss, and no adverse effects on the cotton fabric breaking strength were observed (Abidi and Hequet 2005).

Abidi and Hequet used microwave plasma to impart water repellency to cotton fabric (Abidi and Hequet 2004). Cotton fabric was exposed to Ar-plasma for a specified period of time (240 s). This initial treatment cleans the surface from oils and contaminants and creates radicals on the surface of cellulose. Plasma-treated fabric was then immersed in a vinyl laurate monomer ($CH_3(CH_2)_{10}COOCH{=}CH_2$) solution in xylene. Treated fabrics were then dried and exposed again to Ar-plasma. Fourier transform infrared spectroscopy was used to investigate the grafting efficiency of vinyl laurate monomer on the surface of the cotton fabric. The presence of the vibrations located at 2923, 2855, and 1735 cm^{-1} in the spectra of the treated fabric illustrated the successful grafting of vinyl laurate induced by microwave plasma (Abidi and Hequet 2004). It was reported that the concentration of vinyl laurate should be below 0.664 mol/L for maximum grafting efficiency. This indicated that the use of plasma for monomer grafting on textile substrate should be optimized. The result of the water repellency test showed a water contact angle of 125°, which indicated a hydrophobic cotton fabric surface. The durability to home laundering and tumble drying indicated no effect up to ten laundering cycles (Abidi and Hequet 2004).

Cabrales and Abidi used microwave plasma to graft oleic acid ($CH_3(CH_2)_7CH{=}CH(CH_2)_7COOH$) on the cotton fabric surface (Cabrales and Abidi 2012). Cotton fabric was first exposed to Ar microwave plasma and then immersed in oleic acid solution in xylene. Fourier transform infrared spectroscopy was used to determine the efficiency of grafting, and the water repellency of grafted fabric was assessed by measuring the water contact angle.

5.3.2 Molecular Vapor Modification

Chemical vapor deposition (CVD) represents a chemical reaction process which is conducted in a vapor phase. This process allows the deposition on a surface of atoms, molecules, or both. This technique was used back in 1880 to coat filaments in incandescent lamps with carbon and metal (Pierson 1999). CVD was used in the past to make hydrophobic surfaces (Nakajima et al. 2001). Atomic layer deposition (ALD) is a subcategory of CVD. ALD leads to low and precise thickness of the deposited material (angstrom scale) as well as a high uniformity of the coating. Furthermore, because the reaction occurs in a gas phase, coating can also be deposited in the pores of the substrate surface (Pierson 1999). Aluminum oxide can be deposited onto a surface by means of ALD process. In this process, trimethylaluminum/water binary chemicals are used in the CVD and ALD processes to add additional layers through a sequential reaction (Dillon et al. 1995).

CVD technique is sometimes referred to as nanoparticle vapor deposition (NVD). NVD is similar to ALD with only one exception. In NVD, two binary chemicals are injected at the same time or when the other chemical is present. However, in ALD, each chemical is added to the reaction chamber separately.

NVD and MVD were used to prepare super hydro-/oleophobic cotton fabrics (Abidi et al. 2012; Aminayi and Abidi 2013). Cotton fabric surface was first roughened using trimethylaluminum/water nanoparticles followed by functionalization with (tridecafluoro-1,1,2,2-tetrahydrooctyl)trichlorosilane. Measurements of water contact angles indicated that this process could impart unique hydro-/oleophobic properties (contact angle >160°). The chemicals used to create this surface as well as the thickness of the coatings produced could be controlled. This leads to the generation of minimum waste of reactants as well as preserving the beneficial properties of the fabric.

Self-cleaned textiles have gained interest recently. Self-cleaning property can be imparted either by functionalizing cotton fabric to impart super hydrophobicity (contact angles >150°) (Aminayi and Abidi 2015) or by imparting photocatalytic properties (Abidi et al. 2009). Aminayi and Abidi used nanoparticle vapor deposition of aluminum oxide to create a rough surface followed by molecular vapor deposition of (tridecafluoro-1,1,2,2-tetrahydrooctyl) trichlorosilane (Fig. 5.2) (Aminayi and Abidi 2013).

The advancing contact angle, after the liquid drop advances on the surface, is in general higher than the receding contact angle (after the droplet recedes from a previously wetted surface). The difference between the advancing and receding

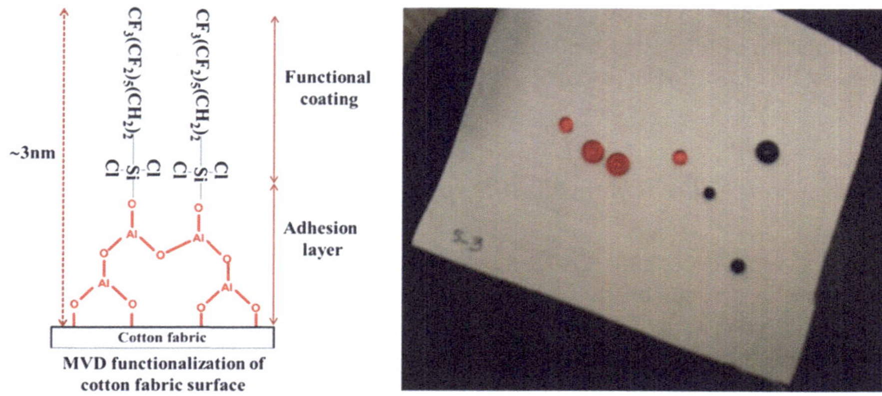

Fig. 5.2 Functionalization of cotton fabric to impart super hydrophobic/oleophobic properties for self-cleaning

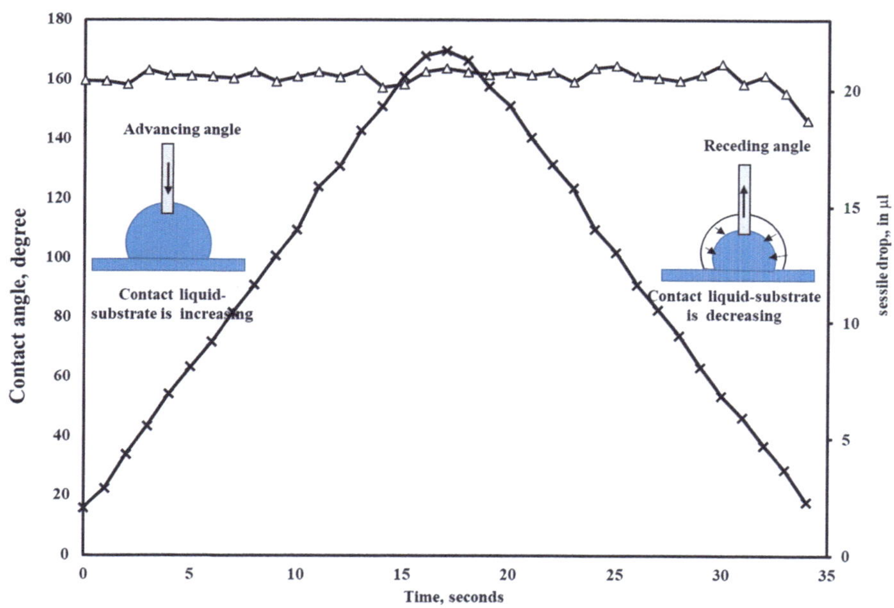

Fig. 5.3 Advancing and receding contact angles

contact angle values is called "hysteresis." This value can be described as the change in the adsorption of liquid on a solid surface as a consequence of the change in the surface energy or the "roughness of the surface." Figure 5.3 shows the advancing and receding contact angles of water on cotton fabric after a treatment consisting of N_2-plasma followed by MVD deposition of trimethylaluminum nanoparticles, then MVD deposition of a bifunctional trichlorosilane blend, and finally MVD deposition of a (tridecafluoro-1,1,2,2-tetrahydrooctyl)trichlorosilane layer (FOTS).

5.4 Antimicrobial Properties

To impart antimicrobial properties, cotton fabrics have been treated with polyethylene glycol (Vigo and Leonas 1999), zirconium complexes (Morris et al. 1981), metallic and organometallic salts (Morris et al. 1981), and natural and inorganic substances such as tea extracts, chitosan, copper, zinc, etc. (Jeon et al. 2003). Silver is widely used as antimicrobial agent. Silver/sodium carboxymethyl cotton for burn dressing was developed by partial exchange of Na^+ by Ag^+ (achieved by treating sodium carboxymethyl nonwoven cotton gauze with $AgNO_3$) (Parikh et al. 2005). The most promising approaches to impart antimicrobial properties are the sol-gel process and cyclodextrins treatment.

5.4.1 *Sol-Gel Process*

Tarimala et al. used the sol-gel process to impart antimicrobial properties to cotton fabric (Tarimala et al. 2006). The authors used dodecanethiol-capped silver nanoparticles (prepared from silver nitrate, tetraoctylammonium bromide, chloroform, sodium borohydride, and 1-dodecanethiol), tetraethyl orthosilicate, ethanol, and water. Cotton fabric was dipped into the solution, soaked for 5 min, and padded at a speed of 4 yd min^{-1} at 2.76 10^5 Pa. The treated fabrics were dried at 60 °C for 10 min and cured at 150 °C for 5 min. Scanning electron microscopy showed the formation of homogenous and uniform coating on the fiber surface. The measurements of antimicrobial properties against *Escherichia coli* were performed by measuring the optical density of the medium containing the bacteria at 600 nm. Treated fabric showed excellent performance against *Escherichia coli.*

5.4.2 *Cyclodextrin Modification*

In textile applications, cyclodextrins have been used to impart various functional properties such as UV protection (El-Tahlawy et al. 2007; Scalia et al. 2006a, b), slow release of fragrances (Martel et al. 2002), insecticide delivery (Romi et al. 2005), and antibacterial properties (Wang and Cai 2008). Triclosan (5-chloro-2-(2,4-dichlorophenoxy)phenol) is a broad-spectrum antibacterial and antifungal agent. It is widely used in consumer products, and its reaction with β-cyclodextrin has been reported in previous studies (Lu et al. 2001; Loftsson et al. 2005).

Several studies reported on the use of monochlorotriazinyl-β-cyclodextrin (MCT-β-CD) to modify textiles (Rehmann et al. 2003; Scalia et al. 2006b; Romi et al. 2005; Wang and Cai 2008). Monochlorotriazinyl functional groups form covalent bonds with hydroxyl groups in cellulose (Fig. 5.4). Hydroxyl groups in cellulose serve as reactive anchors for reactive dyes (Reuscher and Hirsenkorn 1996). MCT-β-CD is the first reactive cyclodextrin derivative produced industrially for textile applications (Szejtli 1997).

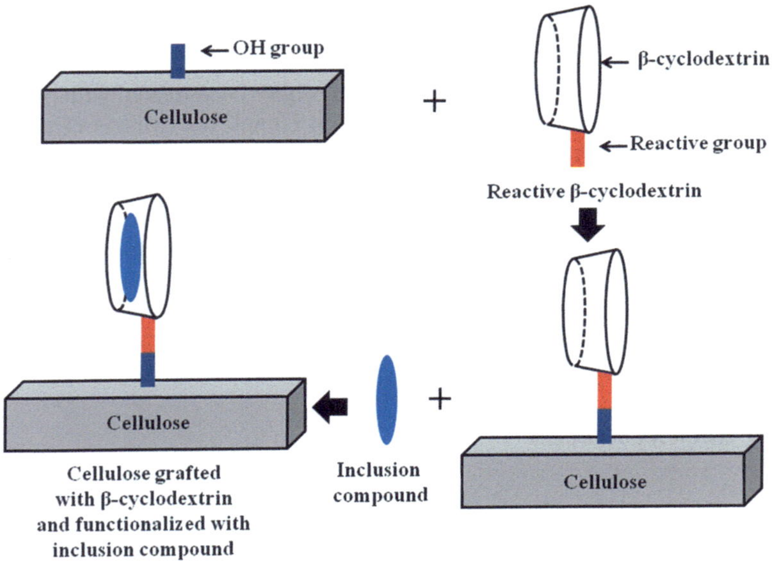

Fig. 5.4 Schematic illustration of the grafting of β-cyclodextrins on cellulose and inclusion of guest compounds in the cavities

MCT-β-CD was grafted to cotton fabrics (Cabrales et al. 2012). MCT-β-CD was stirred for 5 min, and then sodium carbonate (Na_2CO_3) was added to the solution. The amount of Na_2CO_3 was ¼ of the amount of MCT-β-CD. The solution was stirred for 5 min, and the pH of the solution was maintained at 11.5. Cotton fabrics were immersed into MCT-β-CD solution for 5 min and passed through a two-roller laboratory padder (BTM 6-20-190) at a speed of 365 cm/min and an air pressure of 41 kPa. Treated fabrics were dried for 10 min at 50 °C and cured at 150 °C for 10 min in a Ben Dry-Cure Thermosol Oven. Samples were rinsed with water and dried. Figure 5.5 shows an illustration of the chemical reaction between MCT-β-CD and cellulose.

To impart antimicrobial properties, triclosan inclusion in the cavities of cyclodextrins was performed. MCT-β-CD-grafted cotton samples were placed in triclosan solution in ethanol (0.01 mol L^{-1}) for 24 h. Figure 5.6 shows the inclusion of triclosan inside MCT-β-CD cavities. Treated cotton fabrics were dried for 30 min at 40 °C, rinsed with distilled water and with 50% ethanol-water solution, and then dried for 30 min at 40 °C. The 50% ethanolic solution was used due to the low solubility of triclosan in water. Triclosan molecules can be adsorbed also on the fabric surface but not forming inclusion compounds with cyclodextrins. These molecules could interfere with the spectroscopic quantification of the inclusion compounds. Rinsing with alcoholic solutions has been reported to remove adsorbed triclosan molecules on the textile surfaces (Scalia et al. 2006b). Thus, MCT-β-CD-grafted cotton samples were placed in a beaker containing 10 mL of ethanol for 24 h.

To assess the antimicrobial efficiency, colony-forming unit (CFU) of *S. aureus* was measured (Tran et al. 2009; Cabrales et al. 2012). Figure 5.7 shows the CFU of

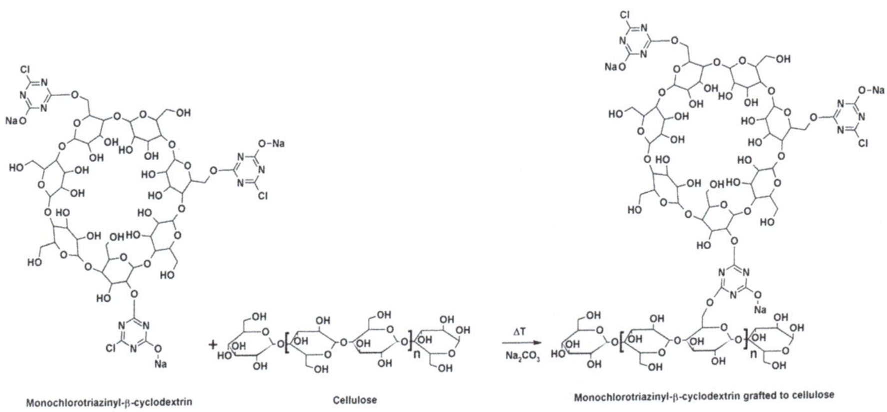

Fig. 5.5 Chemical reaction between monochlorotriazinyl-β-cyclodextrin and cellulose

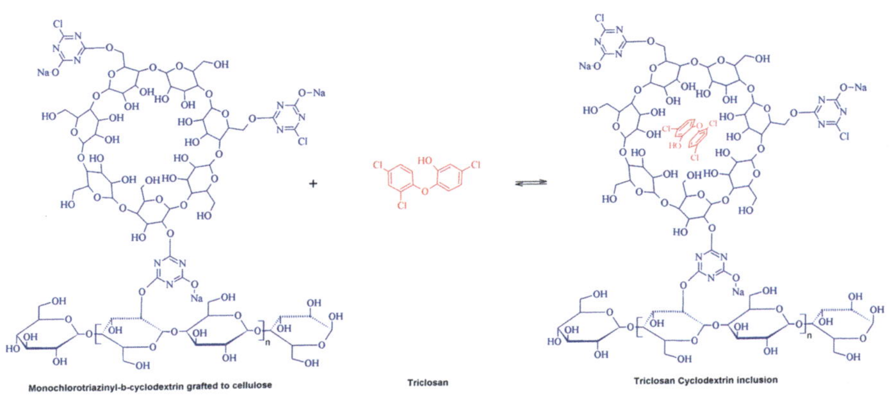

Fig. 5.6 Inclusion of triclosan inside MCT-β-CD cavities

S. aureus of untreated control fabric and MCT-β-CD-grafted cotton fabrics. MCT-β-CD-treated cotton fabric loaded with triclosan molecules exhibited excellent antimicrobial activity against *S. aureus* (100% inhibition efficiency). Upon rinsing the MCT-β-CD-grafted cotton with ethanol, the fabric did not exhibit its antimicrobial property. This indicated that guest compounds in MCT-β-CD cavities can be removed, allowing other guest compounds to be incorporated.

5.5 UV Protection

Ultraviolet radiation (UVR) is composed of UV-A (ranging between 320 nm and 400 nm), UV-B (ranging between 290 nm and 320 nm), and UV-C (ranging between 100 nm and 290 nm). UV-C is absorbed by the ozone layer; however, UV-A and

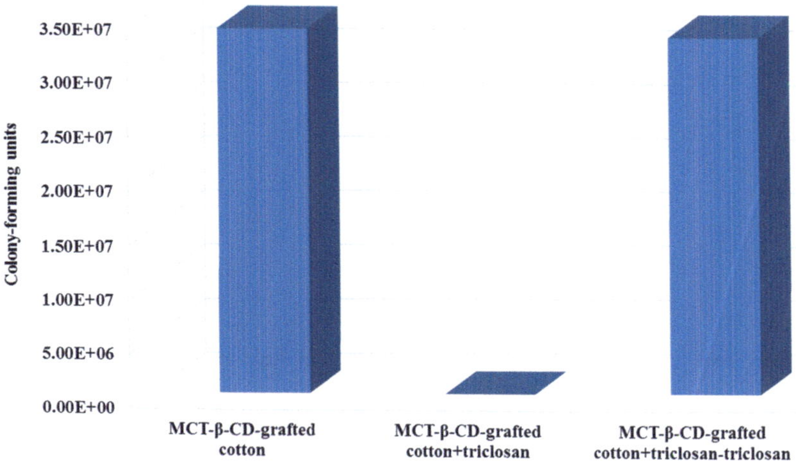

Fig. 5.7 Antibacterial activity test of MCT-β-CD-grafted cotton fabric, MCT-β-CD-grafted cotton fabric with triclosan included in cyclodextrins cavities, and MCT-β-CD-grafted cotton fabric with triclosan included in cavities and then washed with ethanol

UV-B reach the earth surface and can cause various health issues such as skin cancer, sunburn, and photoaging. Special focus has been on UVR transmission of textiles because of the growing demand in the marketplace for lightweight apparel that offers protection from UVR while fostering comfort.

Srinivasan et al. studied the characteristics of the dye and the UV protection provided by the dyed fabrics (Srinivasan and Gatewood 2000). The authors reported that color is not a reliable indicator of the UV protection of the dyed fabrics. Zhou et al. showed that when optical brighteners were added to laundering detergents, the UVR-blocking ability of cotton fabrics and cotton/polyester blends is improved (Zhou and Crews 1998). Eckhardt and Rohwer investigated the factors affecting the UVR transmission of undyed fabrics (Eckhardt and Rohwer 2000). It was concluded that the single best predictor of the ability of undyed woven fabrics to pass the UV-damaging wavelengths was the fabric porosity.

The effect of the concentration of the UV absorber (Rayosan™) was investigated. Scoured and bleached 100% cotton fabrics (typical lightweight summer clothes) were treated using exhaustion method. Figure 5.8 shows the evolution of the UPF as a function of the concentration of the UV absorber. The nonlinear relationship shows that high UV-absorber concentrations result in higher UPF values until a plateau is reached above 8%. The plateau is explained by the saturation of sites (cellulosic OH groups) available for chemical reaction. When the percentage of the UV absorber is greater than 5%, fabric A provides excellent protection from UVR (UPF $\geq$ 40), while fabric B provides only a good protection. This demonstrates that the type of the substrate and the concentration of the UV absorber impact the level of protection provided by the cotton fabric.

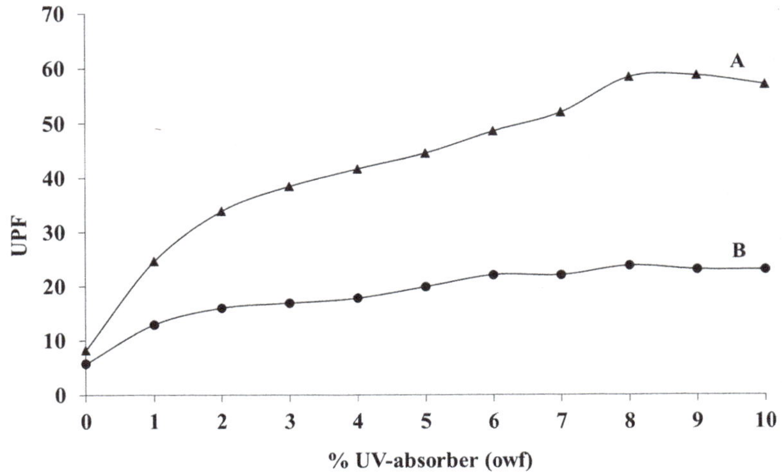

Fig. 5.8 Evolution of the ultraviolet protection factor as a function of UV-absorber concentration. Fabric A: 22.8 Ne, 4.15 oz./yd^2, Fabric B: 40.4 Ne, 3.32 oz./yd^2

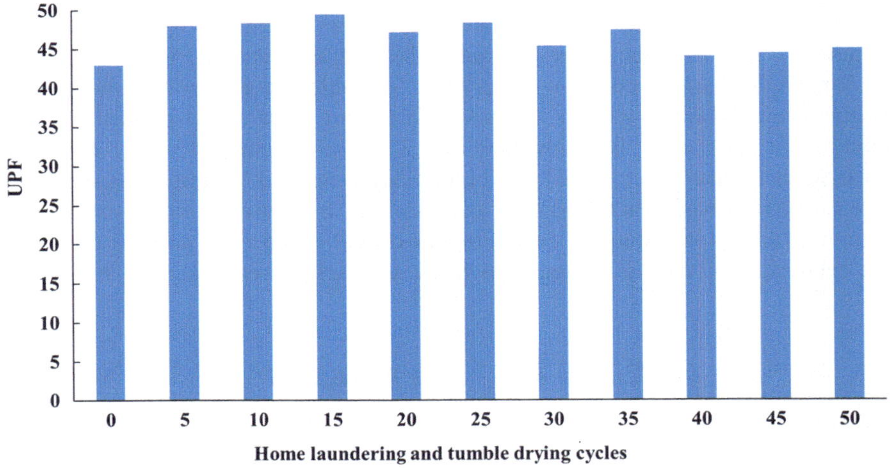

Fig. 5.9 Woven cotton fabric treated with UV absorber: UPF vs. washing cycles

The durability of the treatment was tested after 40 cycles of laundering and tumble drying. Figure 5.9 shows that repeated laundering and tumble drying have no effect on the level of the UVR protection. There is actually a slight increase after 40 cycles, which is attributed to optical brighteners present in commercial detergent used during laundering. The durability of the treatment is attributed to the chemical constitution of this UV absorber. It is a "colorless" reactive dye, formulated from chlorotriazine functional groups, which reacts with cellulose hydroxyl groups through nucleophilic substitution and establishes covalent bonds with cellulose.

5.6 Wrinkle Recovery

Cotton fibers are composed of cellulose macromolecules with repeating anhydroglucose units. The degree of polymerization (number of anydroglucose units) can very between 8000 and 15,000 depending on the cotton cultivar. On each anydroglucose unit, there are three available hydroxyl groups (–OH), which serve as sites for water molecule absorption by establishing many hydrogen bonds with cellulose macromolecules (Fig. 5.10).

When subjected to a deformation, fabrics made of cotton have the tendency to wrinkle. In this process, hydrogen bonds between cellulosic chains in the amorphous regions of fibers break, which allows the chains to slip past one another. Because of the presence of many hydroxyl groups, the hydrogen bonds then reform in new places holding creases and wrinkles in the fiber and fabric. By preventing or restricting cellulosic chain slippage, cotton fabric can resist to the deformation, and wrinkles are not formed or minimized (Perkins 1996). The fabric appears smooth and free of surface deformation. Replacing weak hydrogen bonding by strong covalent bonding in the amorphous regions of the fibers, using appropriate chemical treatment, can restrict the chain slippage of the cellulosic chains (Fig. 5.11).

N-methylol-based products with very low formaldehyde release have been used by the textile industry for many years as cross-linking agent. Other promising compounds are polycarboxylic acids such as 1,2,3,4-butanetetracarboxylic acids and citric acids (Dehabadi et al. 2013). The traditional method for imparting wrinkle-free property is based on the pad-dry-cure process. It consists of impregnating the sample in an aqueous solution containing the cross-linking agent and the appropriate catalyst, padding the impregnated fabric to reach a wet pickup of 90–100%,

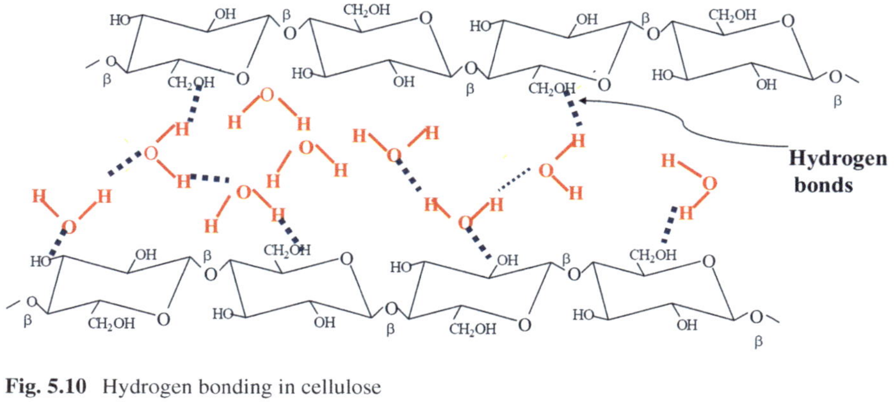

Fig. 5.10 Hydrogen bonding in cellulose

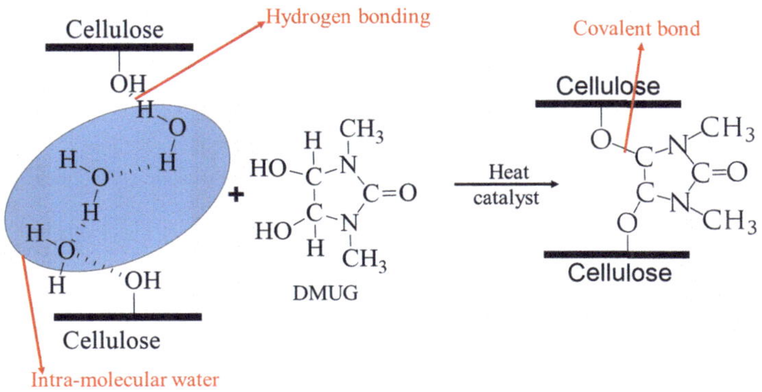

Fig. 5.11 Chemical cross-linking of cellulose chains in cotton fabric to prevent the appearance of wrinkles

drying at 100 °C to remove water, and then curing at 150 °C to establish covalent links between cellulose chains and the cross-linking agent.

The term durable press refers to the smoothness of the fabric. This term is used to describe textile apparels which possess certain wrinkle-resistance properties during daily wear and which require little or no ironing after home laundering and tumble drying.

The American Association of Textile Chemists and Colorists (AATCC) Test Method (TM) 124 outlines the protocol for ascertaining a smoothness grade to a fabric. This test is designed to evaluate the smoothness appearance of fabric specimens after five cycles of repeated home laundering and tumble drying. In general, three trained observers visually evaluate the appearance of the cross-linked fabrics by laying the specimen on a solid surface standing at an incline of 5° from the vertical under specified lighting conditions. The specimen is then compared to six standard replicas (3-D plastic models) showing varying degrees of smoothness and having grades 1 (very wrinkly), 2, 3, 3.5, 4, and 5 (very smooth). The specimen is assigned the grade of the replica it most closely resembles.

Abidi et al. reported on the cross-linking of cotton fabric with DMUG (Abidi et al. 2005). The authors reported an increase in the durable press and wrinkle recovery angle (Fig. 5.12) and drastic decrease in abrasion resistance and breaking force (Fig. 5.13). Overall, there is 50% decrease in breaking strength of the treated fabric and 90% decrease of the abrasion resistance. Cellulose cross-linking in cotton fabric results in stiff material, which can break easily. Furthermore, because chemical cross-linking occurs between OH groups of cellulose and the cross-linking agent, drastic decrease is also observed in the amount of adsorbed water and amount of fixed dye molecules. Wrinkle-free-treated cotton fabrics are often white or light shade color.

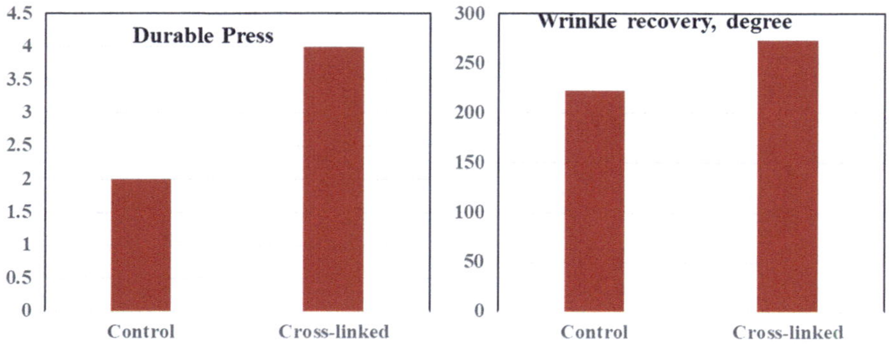

Fig. 5.12 Durable press and wrinkle recovery angle (sum of the warp and filling) of the control and cross-linked cotton fabric

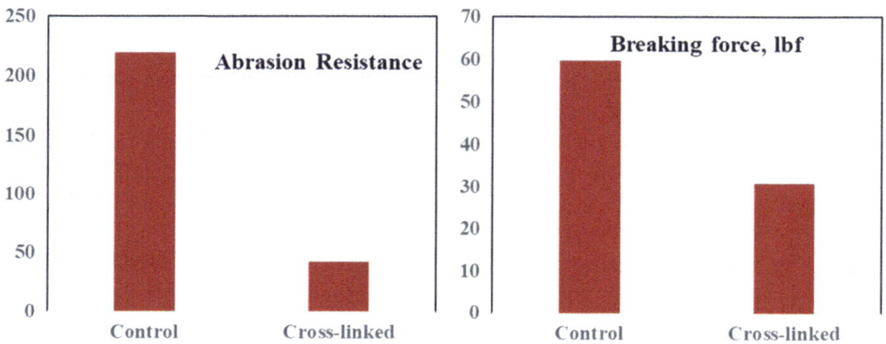

Fig. 5.13 Flex abrasion resistance (in cycles) and breaking for of the control and cross-linked cotton fabric

5.7 Water Repellency Combined With Wrinkle-Free Treatment

Abidi et al. reported that cotton fabric can be easily functionalized with vinyltrimethoxysilane ($H_2C{=}CH{-}Si(OCH_3)_3$) to impart water repellency and wrinkle recovery (Abidi et al. 2007). Functionalization with vinyltrimethoxysilane introduces reactive surface vinyl groups ($-CH{=}CH_2$), which can be used to graft monomers through addition polymerization (Fig. 5.14). The presence of vinyl groups on the surface of the fabric was confirmed by universal attenuated total reflectance FTIR (UATR-FTIR) spectrum of the treated fabric (vibrations 1410 cm^{-1} and 1600 cm^{-1} assigned to C=C stretch and the vibration 756 cm^{-1} assigned to Si-O-Si symmetric stretch).

Figure 5.15 shows the water contact angle and wrinkle recovery angle values as a function of the concentration of vinyltrimethoxysilane. For low concentration of VTMS (<1 mol/L), the treated fabric is hydrophilic. However, the increase of the

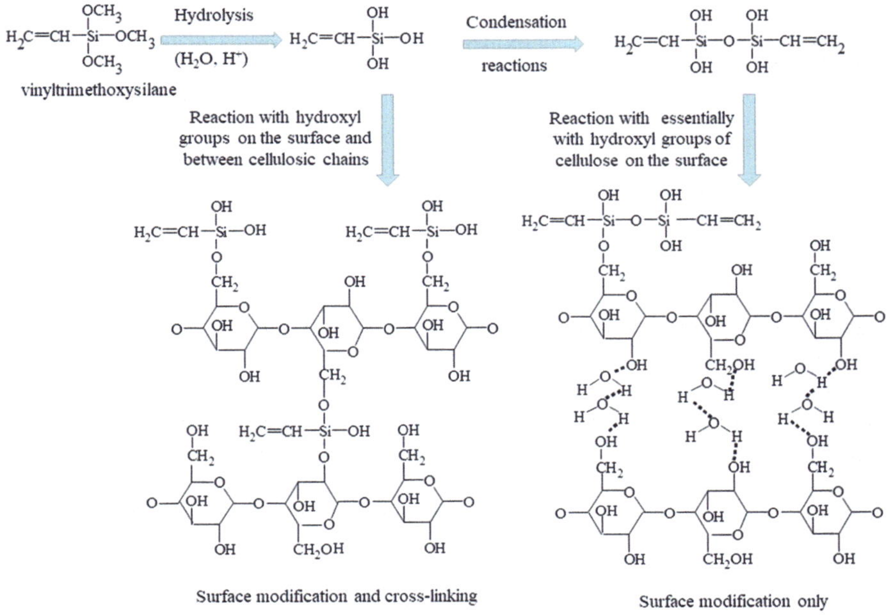

Fig. 5.14 Functionalization of cotton cellulose with vinyltrimethoxysilane (Adapted from (Abidi et al. 2007))

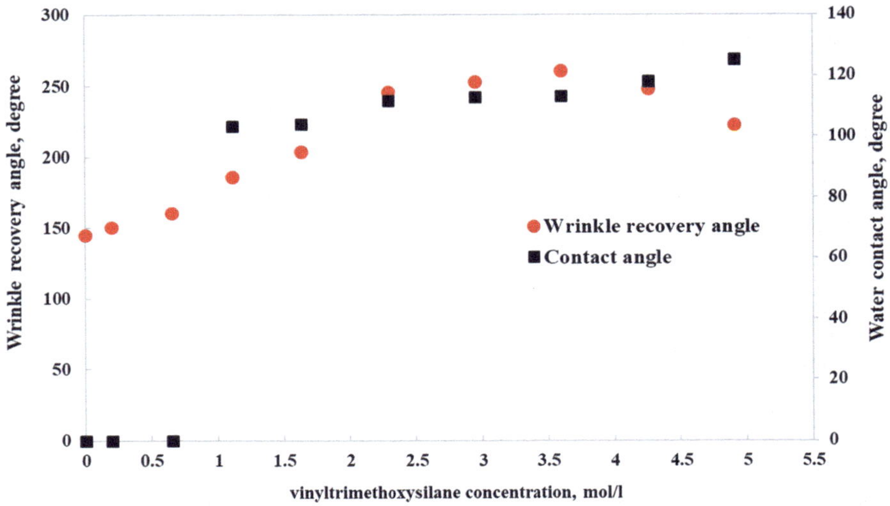

Fig. 5.15 Effects of vinyltrimethoxysilane concentration on the wrinkle recovery and water contact angle

concentration of VTMS increases the contact angles as well as the wrinkle recovery angle values (Fig. 5.15). This demonstrated that the coverage of the surface with (Si-O-Si chains) as well as cross-linking of cellulosic chains occurs.

5.8 Other Finishes

5.8.1 Softening Finish

During the preparation of cotton fabrics for dyeing and finishing, natural waxes are removed by means of scouring and bleaching processes (Schindler and Hauser 2004). Textiles can become embrittled. Treatment with softeners can overcome this deficiency and impart softness. Softening finishes are among the most important of textile chemical after treatment. Applying chemical softeners to textiles imparts soft hand, some smoothness, flexibility and better drape, and pliability. Softeners provide their main effects on the surface of the fibers. Small softener molecules penetrate between fibers and act as internal plasticizer. The physical arrangement of the softener molecules on the fiber surface is important and depends on the ionic nature of the softener molecule and the relative hydrophobicity of the fiber surface. Cationic softeners (e.g., $R_1R_2N(CH_3)_2^+Cl^-$) orient themselves with their positively charged ends toward the partially negatively charged fiber, creating a new surface of hydrophobic carbon chains that provide excellent softening and lubricity seen with cationic softeners. However, anionic softeners (R_1–O–SO_3^- Na^+) orient themselves with their negatively charged ends repelled away from the negatively charged fiber surface, which leads to higher hydrophilicity but less softening than with cationic softeners.

Cationic softeners provide the best softness and are reasonably durable to laundering. They can be applied using exhaustion method to all fibers from a high liquor to goods ratio. The finished textile products acquire hydrophobic surface. Anionic softeners are generally stable to heat at normal textile processing temperatures and compatible with other components of dye and bleach baths.

5.8.2 Flame-Retardant Finishes

The resistance of textile apparels to fire is imparted to textiles using flame-retardant finishes (Schindler and Hauser 2004). Protecting consumers, firefighters, and emergency personnel requires protection from flames. Floor coverings, mattresses, furniture, and drapery also require flame-retardant treatment, especially when used in public buildings. Military and airline industries have multiple needs for flame-retardant textiles. Commercially efficient flame-retardant agents should have little or no adverse effect on the textiles' physical and aesthetic properties, should be produced by a simple process with conventional equipment and inexpensive chemicals, and should be durable to repeated home laundering/tumble drying and dry cleaning.

The mechanism of combustion should be explained in order to understand the mechanisms of effective flame retardants. Combustion is an exothermic process that requires three components: Source of heat, source of oxygen, and suitable fuel. Combustion can be self-catalyzing and can continue as long as oxygen, fuel supply, or excess heat is available. When textiles are exposed to heat, the temperature of fibers increases until the pyrolysis temperature (Tp) is reached. At this temperature, fibers are subjected to irreversible chemical changes leading to the production of nonflammable gases, char, liquid condensates, and flammable gases. If the temperature continues to increase until it reaches the temperature of combustion (Tc), flammable gases are generated. The combustion process can generate additional heat needed to continue the pyrolysis. Therefore, flame-retardant chemicals applied to textile should disrupt this cycle. Several approaches have been proposed (Schindler and Hauser 2004). Providing a heat sink on or in the fiber by using materials that, when subjected to heat, thermally decompose through strongly endothermic reactions. When enough heat is absorbed by these reactions, the pyrolysis temperature of the fiber is not reached, and no combustion takes place. Examples of these chemical compounds include the use of aluminum hydroxide three hydrate ($Al_2O_3\ 3H_2O$) and calcium carbonate ($CaCO_3$) as fillers in polymers and coatings. When subjected to heat, $Al_2O_3\ 3H_2O$ changes to Al_2O_3 and $3H_2O$, while $CaCO_3$ changes to CaO and CO_2. Another approach consists of applying a material (such as boric acid and its hydrate salts) which is able to form an insulating layer around the fiber at temperatures below the pyrolysis temperature of the fiber. When subjected to heat, treated textiles with boric acid and its hydrate salts, boric acid-based compounds release water vapor and produce a foamed glassy surface on the fiber, insulating the fiber from the applied heat and oxygen. Other approaches to achieve flame retardancy consist of influencing the pyrolysis reaction to produce less flammable volatiles and more residual char. This class of chemical compounds is based on phosphorous-containing flame retardants. This class of compound, after they produce phosphoric acid through thermal decomposition, cross-links with hydroxyl-containing polymers which alter the decomposition and yield less flammable by-products.

The thermal degradation of cellulose fibers leads to the formation of small depolymerization products such as levoglucosan. Levoglucosan volatile pyrolysis products are extremely flammable materials and are the main contributors to cellulose combustion. Therefore, chemical compounds designed to act as flame retardants for cellulose should prevent levoglucosan formation. Phosphoric acid acts as an effective flame-retardant agent by reducing levoglucosan generation and catalyzes the dehydration and the carbonization. Chemical compounds which can lead to phosphoric acid formation during early stages of fiber pyrolysis can act as successful flame retardants for cellulose.

5.8.3 Enzymatic Modification

The use of enzymes for cellulosic textiles treatment can produce permanent effects (Schindler and Hauser 2004). This process is termed as bio-finishing or bio-polishing. This process removes protruding fibers and slubs from the fabric surface, which can significantly affect the appearance of the fabric by reducing pilling and providing softness and smoothness to the fabric. In the denim process, bio-finishing is used to reduce the abrasive stones and the aggressive chlorine chemistry to impart the desired "worn" look that the consumers desire. The bio-finishing process suffers from several disadvantages such as formation of fiber dust, reproducibility of the effect, and loss of strength. The enzymes used for cellulose treatment are high molecular weight proteins having complex three-dimensional structures composed of long chains of amino acids. The enzyme and substrate form "lock and key" complex that requires the enzyme to have a specific molecular alignment in order to act as a catalyst. The enzymes that hydrolyze cellulose are found in nature in both *Trichoderma* and *Humicola* fungi. *Cellulases* can catalytically hydrolyze the β(1–4) linkages between adjacent repeating glucose units in cellulose. Enzymes contain multiple components that work synergistically to yield glucose from cellulose. At least four components have been identified as being important in providing efficient glucose production. Endo-glucanases hydrolyze cellulose at random locations. Beta-glucanases hydrolyze cellulose polymers from the nonreducing end producing glucose and leaving cellulose chain with one less repeating unit. Cellobiohydrolases produce cellobiose. Cellobiases convert cellobiose into glucose.

It is generally accepted that the mechanism of cellulase interactions with cellulose is via first an adsorption of endo-glucanase, beta-glucanase, or cellobiohydrolase components onto the fiber surface followed by complex formation with the cellulose polymer chain and water. The action of enzymes on cellulose is affected by several factors such as pH, temperature, time, and mechanical agitation. Because enzymes are true catalysts and are not consumed during the chemical reaction, cellulose hydrolysis will continue until the conditions of the reaction are changed (change of the pH or temperature) or until cellulose is removed from the solution.

5.9 Conclusions

Cotton fabric finishing is the most important final step to impart the required functional properties that the consumers desire. The term finishing is often broadly used to describe any treatment to improve the quality of the fabric. This chapter reviewed the most important treatment to impart functional properties. Wrinkle-resistance treatment is performed to prevent the formation of wrinkles on the fabric surface after home laundering and tumble drying. Water and oil repellency treatment is performed to limit liquid or oil diffusion inside the fabric. Antimicrobial treatment is performed to protect against the growth of microorganisms. Flame-retardancy treatment is applied to protect consumers from unsafe textiles. UV protection treatment is required to protect sensitive skins from harmful effects of sunlight ultraviolet radiation.

References

Abidi N, Aminayi P, Cabrales L et al (2012) Super-hydrophobic cotton fabric prepared using nanoparticles and molecular vapor deposition methods. In: Functional Materials from Renewable Sources, vol 1107. American Chemical Society, Washington, pp 149–165

Abidi N, Cabrales L, Hequet E (2009) Functionalization of a cotton fabric surface with Titania Nanosols: applications for self-cleaning and UV-protection properties. ACS Appl Mater Interfaces 1:2141–2146

Abidi N, Hequet E (2004) Cotton fabric graft copolymerization using microwave plasma. I. Universal attenuated total reflectance-FTIR study. J Appl Polym Sci 93:145–154

Abidi N, Hequet E (2005) Cotton fabric graft copolymerization using microwave plasma. II. Physical properties. J Appl Polym Sci 98:896–902

Abidi N, Hequet E, Tarimala S (2007) Functionalization of cotton fabric with vinyltrimethoxysilane. Text Res J 77:668–674

Abidi N, Hequet E, Turner C et al (2005) Objective evaluation of durable press treatments and fabric smoothness ratings. Text Res J 75:19–29

Aminayi P, Abidi N (2013) Imparting super hydro/oleophobic properties to cotton fabric by means of molecular and nanoparticles vapor deposition methods. Appl Surf Sci 287:223–231

Aminayi P, Abidi N (2015) Ultra-oleophobic cotton fabric prepared using molecular and nanoparticle vapor deposition methods. Surf Coat Technol 276:636–644

Cabrales L, Abidi N (2012) Microwave plasma induced grafting of oleic acid on cotton fabric surfaces. Appl Surf Sci 258:4636–4641

Cabrales L, Abidi N, Hammond A et al (2012) Cotton fabric functionalization with cyclodextrins. J Mater Environ Sci 3:561–574

Cocinero EJ, Gamblin DP, Davis BG et al (2009) The building blocks of cellulose: the intrinsic conformational structures of cellobiose, its epimer, lactose, and their singly hydrated complexes. J Am Chem Soc 131:11117–11123

Dehabadi VA, Buschmann HJ, Gutmann JS (2013) Durable press finishing of cotton fabrics: an overview. Text Res J 83:1974–1995

Di Mundo R, De Benedictis V, Palumbo F et al (2009) Fluorocarbon plasmas for nanotexturing of polymers: a route to water-repellent antireflective surfaces. Appl Surf Sci 255:5461–5465

Dillon AC, Ott AW, Way JD et al (1995) Surface chemistry of Al_2O_3 deposition using $Al(CH_3)_3$ and H_2O in a binary reaction sequence. Surf Sci 322:230–242

Eckhardt C, Rohwer H (2000) UV protector for cotton fabrics. Text Chem Color Am Dyestuff Rep 32:21–23

El-Tahlawy K, El-Nagar K, Elhendawy AG (2007) Cyclodextrin-4 Hydroxy benzophenone inclusion complex for UV protective cotton fabric. J Text Inst 98:453–462

Eren HA, Anis P, Davulcu A (2009) Enzymatic one-bath desizing - bleaching - dyeing process for cotton fabrics. Text Res J 79:1091–1098

Favia P, Cicala G, Milella A et al (2003) Deposition of super-hydrophobic fluorocarbon coatings in modulated RF glow discharges. Surf Coat Technol 169–170:609–612

Hsieh C-T, Cheng Y-S, Hsu S-M et al (2010) Water and oil repellency of flexible silica-coated polymeric substrates. Appl Surf Sci 256:4867–4872

Jeon HJ, Yi SC, Oh SG (2003) Preparation and antibacterial effects of $Ag-SiO_2$ thin films by sol-gel method. Biomaterials 24:4921–4928

Klemm D, Heublein B, Fink HP et al (2005) Cellulose: fascinating biopolymer and sustainable raw material. Angew Chem Int Ed 44:3358–3393
Kulinich SA, Farzaneh M (2005) On wetting behavior of fluorocarbon coatings with various chemical and roughness characteristics. Vacuum 79:255–264
Lee I, Evans BR, Woodward J (2000) The mechanism of cellulase action on cotton fibers: evidence from atomic force microscopy. Ultramicroscopy 82:213–221
Liu D, Li W, Feng Z et al (2009a) Plasma enhanced CVD of fluorocarbon films by low-pressure dielectric barrier discharge. Surf Coat Technol 203:1231–1236
Liu YH, Wang XK, Luo JB et al (2009b) Fabrication and tribological properties of superhydrophobic surfaces based on porous silicon. Appl Surf Sci 255:9430–9438
Loftsson T, Ossurardottir IB, Thorsteinsson T et al (2005) Cyclodextrin solubilization of the antibacterial agents triclosan and triclocarban: effect of ionization and polymers. J Incl Phenom Macrocycl Chem 52:109–117
Lu J, Hill MA, Hood M et al (2001) Formation of antibiotic, biodegradable polymers by processing with Irgasan DP300R (Triclosan) and its inclusion compound with beta-cyclodextrin. J Appl Polym Sci 82:300–309
Martel B, Morcellet M, Ruffin D et al (2002) Capture and controlled release of fragrances by CD finished textiles. J Incl Phenom Macrocycl Chem 44:439–442
Maxwell JM, Gordon SG, Huson MG (2003) Internal structure of mature and immature cotton fibers revealed by scanning probe microscopy. Text Res J 73:1005–1012
Morris CE, Vigo TL, Welch CM (1981) Binding of organic anti-microbial agents to cotton fabric as zirconium complexes. Text Res J 51:90–96
Bhat NV, Benjamin YN (1999) Surface resistivity behavior of plasma treated and plasma grafted cotton and polyester fabrics. Text Res J 69:38–42
Nakajima A, Hashimoto K, Watanabe T (2001) Recent studies on super-hydrophobic films. Monatsh Chem 132:31–41
Parikh DV, Fink T, Rajasekharan K et al (2005) Antimicrobial silver/sodium carboxymethyl cotton dressings for burn wounds. Text Res J 75:134–138
Park S, Baker JO, Himmel ME et al (2010) Cellulose crystallinity index: measurement techniques and their impact on interpreting cellulase performance. Biotechnol Biofuels 3:10
Perkins WS (1996) Textile coloration and finishing. Carolina Academic, Durham
Peterlin A, Ingram P (1970) Morphology of secondary wall fibrils in cotton. Text Res J 40:345
Pierson HO (1999) Handbook of chemical vapor depostion [ie deposition] (CVD): principles, technology, and applications. William Andrew, New York
Rehmann L, Yoshii H, Furuta T (2003) Characteristics of modified beta-cyclodextrin bound to cellulose powder. Starch-Starke 55:313–318
Reuscher H, Hirsenkorn R (1996) BETA W7 MCT - new ways in surface modification. J Incl Phenom Mol Recognit Chem 25:191–196
Rivlin J (1992) The dyeing of textile fibers theory and practice. Joseph Rivlin, Philadelphia
Romi R, Lo Nostro P, Bocci E et al (2005) Bioengineering of a cellulosic fabric for insecticide delivery via grafted cyclodextrin. Biotechnol Prog 21:1724–1730
Scalia S, Tursilli R, Bianchi A et al (2006a) Incorporation of the sunscreen agent, octyl methoxycinnamate in a cellulosic fabric grafted with beta-cyclodextrin. Int J Pharm 308:155–159
Scalia S, Tursilli R, Sala N et al (2006b) Encapsulation in lipospheres of the complex between butyl methoxydibenzoylmethane and hydroxypropyl-beta-cyclodextrin. Int J Pharm 320:79–85
Schindler WD, Hauser PJ (2004) Chemical finishing of textiles. Woodhead, Cambridge
Spicka N, Tavcer PF (2013) Complete enzymatic pre-treatment of cotton fabric with incorporated bleach activator. Text Res J 83:566–573
Srinivasan M, Gatewood BM (2000) Relationship of dye characteristics to UV protection provided by cotton fabric. Text Chem Color Am Dyestuff Rep 32:36–43
Szejtli J (1997) Utilization of cyclodextrins in industrial products and processes. J Mater Chem 7:575–587
Tarimala S, Kothari N, Abidi N et al (2006) New approach to antibacterial treatment of cotton fabric with silver nanoparticle-doped silica using sol-gel process. J Appl Polym Sci 101:2938–2943

Tran PL, Hammond AA, Mosley T et al (2009) Organoselenium coating on cellulose inhibits the formation of biofilms by Pseudomonas aeruginosa and Staphylococcus aureus. Appl Environ Microbiol 75:3586–3592

Vigo TL, Leonas KK (1999) Antimicrobial activity of fabrics containing crosslinked polyethylene glycols. Text Chem Color Am Dyestuff Rep 1:42–46

Wang JH, Cai ZS (2008) Incorporation of the antibacterial agent, miconazole nitrate into a cellulosic fabric grafted with beta-cyclodextrin. Carbohydr Polym 72:695–700

Zhou Y, Crews PC (1998) Effect of OBAs and repeated launderings on UVR transmission through fabrics. Text Chem Color 30:19–24

Chapter 6
Color Cotton and Its Utilization in China

Wenfang Gong, Xiongming Du, Yinhua Jia, and Zhaoe Pan

6.1 Introduction

Naturally colored cottons (NCCs) refer to the fibers that have natural colors other than white. NCC fibers exist in various hues including light to dark brown, red, rust, and green and are found among both tetraploid and diploid *Gossypium* species (Hinchliffe et al. 2016). NCCs are typically grown as a source of fiber for niche textile markets that promote the use of natural colors in textiles as an alternative to dying scoured and bleached cotton fibers. This use has great merit considering that global textile processing generates one of the largest toxic chemical waste streams with negative environmental impacts. Fabrics and clothes made from NCC fibers leave smaller footprints to the environment, and therefore, NCC is often called "ecologically friendly cotton" (Hu 2004).

Currently the NCC fibers include two large groups: brown and green (Fig. 6.1). Under each group, the fibers can be classified into subgroups based on the color difference. For example, for brown cottons, there are light brown, reddish brown, coffee brown, dark brown, etc. At present, the naming of color cottons is complex and often confusing. It would be helpful if cotton community can provide a standard nomenclature to name color cottons based on a set of predefined criteria. The NCCs are present in all four cultivated *Gossypium* species with the most variations found in upland cotton (*G. hirsutum*), the second in *G. arboreum*, and few in *G. barbadense* and *G. herbaceum*. NCCs are grown in 27 countries with China and India being the leading producers.

The pigments present in NCC fibers are often not stable. They are greatly affected by the environmental factors especially by the weather conditions during the period

W. Gong · X. Du (✉) · Y. Jia · Z. Pan
Institute of Cotton Research, Chinese Academy of Agricultural Sciences (ICR, CAAS)/State Key Laboratory of Cotton Biology, Anyang, Henan, China
e-mail: duxiongming@caas.cn

D. D. Fang (ed.), *Cotton Fiber: Physics, Chemistry and Biology*,
https://doi.org/10.1007/978-3-030-00871-0_6

Fig. 6.1 Different color cotton bolls. (**a**) Dark brown, (**b**) light brown, (**c**) dark green, (**d**) light green, (**e**) white normal cotton

of fiber development (from flower opening to boll maturation). After a cotton boll opens, sunlight could make green cotton fibers less green and, on the other hand, make the brown fibers darker. Studies have shown that shades greatly affected both the quality and color of NCC fibers. Shading cotton plants at 25 days post anthesis could make brown cotton lighter while make the green cotton fiber greener as compared to the controls (Pan et al. 2006). The instability of the NCC fiber pigments mainly falls into two categories. First, the color of NCC fibers fades or disappears completely after exposure to sunlight. Second, the offsprings of the NCC segregate as color, white, and mixture of color and white. Among the dark brown, brown, and light brown cottons, the color of dark brown fiber is usually stable. The pigment stability of brown cottons differs greatly between genotypes. The brown cotton color is usually more stable in *G. arboreum* than in *G. hirsutum*. In *G. hirsutum*, a great difference exists in terms of pigment stability. In some lines the fiber colors are stable, while in other lines fiber colors can degrade to gray or grayish white after exposure to sunlight. The light brown types are usually not genetically stable and often segregate many types in progeny plants. Green cottons are very unstable and fade rapidly. It is a common phenomenon to see a contrasting color difference between green cottons stored under dark and those stored under nature light. The pigment instability of color cottons poses a big challenge and difficulty for large-scale storage and utilization of naturally colored cotton fibers.

Naturally colored cottons have many disadvantages as compared to the white cottons. A NCC usually has lower gin turnout. Its fibers are shorter, weaker, and less mature than white cottons. For brown cotton fibers, their upper half mean length is less than 22 mm, strength about 15 cN/tex. A dark brown cotton is even worse in both fiber quality and yield than a brown or light brown cotton and usually matures very late. For green cottons, their fiber upper half mean length is about 25 mm, strength 11–18 cN/tex, and very low micronaire (MIC) value (about 2–2.5).

The NCC has been existing probably since the dawn of cotton genus formation. Archeological evidence showed that Peruvians used color cottons to make quilts about 2500 years ago. Columbus and expediters after him discovered color cottons in Central and South American countries. However, the color cottons were not planted due to very low fiber quality that did not meet the textile requirement. Breeding and improvement of color cottons as a crop did not start until the late 1960s. As of today, color cottons are grown in 27 countries including China, India, the USA, Pakistan, Russia, Turkey, etc.

6.2 Current Status of Color Cotton Research and Utilization

6.2.1 Color Cotton Research and Utilization in China

China has a long history to utilize color cottons. As back as 2000 years ago, Chinese used color cottons (*G. arboreum*) in their daily life. Beginning Qing dynasty (1644 AD), high demand for red-purple fabrics facilitated planting of color cottons in lower range of Yangtze River. All color cottons planted then were diploid *G. arboreum*. During the period of 1784–1833, China exported more than 40 million pieces of handwoven fabrics to Europe and America, and most of them were from color cottons (specifically, brown Asia cotton). In the year 1819 alone, about 3.3 million pieces of red-purple cotton fabrics produced in Shanghai, Zhejiang, and Jiangsu were exported. In the eighteenth and nineteenth centuries, clothes made from China-produced color cotton fabrics were very popular in Great Britain.

Like the rest of the world, color cotton research in China started in the late 1980s. Since 1987, researchers in the Institute of Cotton Research of the Chinese Academy of Agricultural Sciences (ICR-CAAS) have been conducting a series of research projects to facilitate utilization of color cottons. They used a variety of breeding techniques such as interspecific hybridization and the mutant induction in combination with conventional breeding to break the negative correlations between fiber quality and yield. They released several color cotton varieties that obtained certificates of ecologically friendly plant varieties from Chinese government. The first generation of color cotton lines includes Brown cotton 1, 263, and 3-944 (Fig. 6.2) and Green cotton 1. Later on, researchers in this institute developed several more color cotton varieties that have higher yield and gin turnout. For example, the variety Brown cotton 128 has dark brown fibers, gin turnout as high as 36–40.1%, pre-frost yield rate 97.9%, fiber length 28–29 mm, strength 21–22.9 cN/tex, and MIC value of 4.5. In addition, it produces 20% higher output than that of the first-generation line Brown cotton 1. Using transgenic technology, Chinese researchers transferred pest insect resistance genes into color cottons that could resist to insects such as bollworms. The variety ZMS 51 (CCRI 51) is such an example. This variety is suitable for planting in the Yellow River valley region, and its lint yield could reach 1350 kg/ha. Its fibers are light brown and have an average length of 30 mm

Fig. 6.2 Color cotton variety Brown 263 (**a**) and Color cotton variety CCRI 81 (**b**)

and strength of 30 cN/tex. In cooperation with Xingjiang Naturally Color Cotton Development Company and Chinese Academy of Sciences, the ICR-CAAS established an extensive research program to promote utilization of color cottons. They tested about 100 color cotton accessions in multiple cotton-growing regions in China and selected accessions that are suitable to any a given region. Under the leadership of this institute, the interests of growing color cotton in China have been constantly growing, and currently there are more than ten organizations involving in color cotton research and development. After more than 30 years of research and breeding, Chinese researchers have developed 38 color cotton varieties, which account for 90% of all color cotton varieties registered in the world.

In China, a great majority of color cottons are being grown in Xinjiang and Yellow River region. In 1994 scientists at the ICR-CAAS started to plant color cottons in Yellow River region and in 1995 in Xinjiang province. Color cotton acreage in Xinjiang has been increasing at a rate of 667 ha (10,000 mu) per year (Deng et al. 2010). By 2015, Xinjiang has become the world largest color cotton production region with total acreage of about 13,333 ha (200,000 mu) and lint production of 15,000 metric tons. The average yield of brown cottons was about 1200–1350 kg/ha, about 80% of white cotton average yield. The average yield of green cottons was 1050–1200 kg/ha, about 65% of white cotton average yield (Tan and Zhou 2015). In recent years, the acreage of color cottons in China remains about 15,000 ha per year with small fluctuation based on the market demand. The overall trend is very slowly growing. As the world largest color cotton producer, China has established a comprehensive color cotton production and service chain that includes breeding, seed production, plantation, textile and clothes, and marketing. A few fashion brands such as "Geocolor," "Rainbow," "Docare," "DingGuagua," "Xueyang," etc. have been develope din China.

6.2.2 Color Cotton Research and Utilization in the USA

Color cotton research in the USA started in California in the 1970s. In 1982, Sally V. Fox made single plant-based selections using the color cotton germplasm preserved in the USDA-ARS experimentation station in California. Mrs Fox used the color cotton fibers she produced to make fabrics, towels, bedsheets, and shirts. Her color cotton products were quite popular. She also exported the color cotton fibers to Asian countries at much higher prices than white cotton. It was a business success, and several textile companies funded her research. After several years of breeding, in 1988 she obtained the first color cotton line that had fibers suitable for mechanical spinning. In 1990, two of her color cotton varieties, "COYOTE" (brown) and "GREEN," obtained the US plant variety patent. Afterward, she established Arizona Natural Color Cotton Company, which was the largest company in the USA to produce and market color cotton products. As compared to white cotton, the Fox color cotton fibers were shorter and weaker. To overcome these drawbacks, Fox tried to cross her materials with color cotton germplasm from Peru to improve fiber quality.

Harvey Campbell and Raymond Bird Brothers obtained color cotton germplasm from Texas A&M University; University of California, Berkeley; and USDA-ARS and started color cotton breeding. In 1992, they established the BC Cotton Company in California and put into the commercial production of color cotton fibers. In 1993 they planted 800 hectares of color cotton in three states (California, Texas, and Arizona). However, the US color cotton production disappeared completely in 2000.

6.2.3 Color Cotton Research and Utilization in Russia

Russian color cotton research began in the late 1960s. It was initiated by researchers in the former Soviet Union Academy of Agricultural Sciences. The Institute of Special Cotton located in the southwestern Kiev was the main organization for color cotton research. The weather and climate in Kiev are good for color cotton growth and production. Russian scientists bred cotton varieties with different colors including pale yellow, pink, pale blue, light brown, and gray. The Russian color cotton varieties did not attract much attention due to weak fiber and instability of fiber colors.

6.2.4 Color Cotton Research and Utilization in Central Asian Countries

Central Asian countries include four of the former Soviet Union countries, i.e., Kazakhstan, Tajikistan, Turkmenistan, and Uzbekistan. These countries have state-run agriculture experiment stations responsible for color cotton research. Their research scale is usually small. The pilot research areas are guarded with steel wire nets to protect from human or animal damage and prevent cross-pollination by insects. They have released a few color cotton varieties such as light yellow varieties CPK-1 and CPK-2. The Uzbekistan color cotton research work is mainly run by the Bukhara Province cotton seed corporation. The foundation germplasm was CPK-1 and CPK-2 introduced from Turkmenistan. The color cotton-growing season in Uzbekistan is April to September, and seed cotton yield reaches to 2500–3000 kg/ha.

6.2.5 Color Cotton Research and Utilization in Egypt

Genetic and breeding research of color cotton in Egypt began in mid-1970. The research is led by scientists affiliated to the Egyptian Ministry of Agriculture, and their research base is located in Giza, a suburb of Cairo in Nile Delta. Their research is confidential and rarely opens to outsiders. Scientists involved in the research must be Egyptian citizens. They have bred color cotton varieties with light red, pale yellow, light blue, and gray colors. They also developed a variety with multiple colors in the same plant. An Egyptian-colored cotton variety is usually characterized as big plant, large boll, long, and strong fiber but instability of intended fiber colors. Commercial production of color cotton in Egypt is very small.

6.2.6 Color Cotton Research and Utilization in Peru

As back as 2500 years ago, the Peruvians living in Mochika area planted and used native color cottons. However, wars, natural or man-made disasters, plus ignorant of agriculture business led to the loss of color cotton seeds. In 1988 color cotton seeds were discovered in a Mochika tomb. The seeds were transported to Germany for scientific evaluation and germination test. Surprisingly, these seeds germinated and produced gray, red, and reddish yellow cotton fibers. From this, Peruvians began their color cotton research. In Peru, there are many perennial colored cotton plants, some of them as high as 5 m, each plant producing more than 10 kg of seed cotton. Currently, they have bred six cotton varieties with different colors. In 1992, Peru produced 300 tons of organic color cotton, and most of them were exported to the USA, Europe, Japan, and other countries.

In summary, although color cottons have many shortcomings such as low yield, short and weak fibers, and instability of colors, market demand for nature color cotton is slowly growing. Increased research on color cotton may improve its yield and fiber quality to the comparable level of white cotton.

6.3 Genetics of Color Cotton Traits and Molecular Marker Development

Genetic control of fiber colors was studied by several researchers but with conflicting conclusions. As early as in the 1930s, Ware (1932) crossed color cottons with white cottons and counted the fiber color segregation among the F_2 progeny plants. He observed the segregation ratios as 1:2:1 or 1:1 and determined the fiber color in upland cotton was controlled by one single gene with incomplete dominance. Endrizzi and Taylor (1968) located the *Lc1*, the genetic locus for brown cotton, on chromosome 7 using monosomic techniques. Zhang (1998) reported that fiber color was a quantitative trait and controlled by multiple genes. Zhan et al. (2008) showed that F_2 progeny plants derived from a cross between a brown cotton and a white cotton had dark brown, light brown, and white and the ratio was 1:2:1. He determined that the brown fiber was controlled by one gene with incomplete dominance. A heterozygous plant would have light brown fiber due to the incomplete dominance. He also speculated that besides the major gene, some "modifier genes" would affect brown fiber color inheritance. Zhang et al. (2002) crossed brown cottons and white cottons in 18 different combinations. All F_1 plants had light brown color, and the F_2 plants segregated as dark brown, light brown, and white. The ratio between brown (dark brown and light brown) and white was 3:1. Their results supported the conclusion that the brown fiber color is controlled by one single gene with incomplete dominance. This conclusion has also been confirmed by other studies (Geng et al. 1998; Shi et al. 2002; Li et al. 2004). Shi et al. (2002) indicated that the lint and fuzz fibers of brown cotton were both controlled by one gene with incomplete dominance. However, Sun et al. (2008) suggested a single dominant gene controls the lint and fuzz of brown cotton fibers. Both Sun and Shi demonstrated that the lint and fuzz fiber colors were correlated. The gene controlling fiber color may also affect the length. Feng et al. (2010) found that fuzz was usually darker in color than lint in the brown fiber. This indicates that genes controlling lint and fuzz color development in brown cotton might crosstalk during the fiber development process. Zhang et al. (2009) mapped the *Lc1* locus within an interval of 17.6 cM on the chromosome 7. Two microsatellite markers CIR238 and NAU3181 flanked the *Lc1* locus. Li et al. (2012a) further narrowed the interval to 7.9 cM with NAU4030 and CGR5119 as flanking markers. In a separate research, Wang et al. (2014) also mapped the *Lc1* locus on the chromosome 7 and identified NAU2862 and NAU1043 as flanking markers; however, the genetic distance between these two markers was 11.6 cM. Wen et al. (2018) fine mapped the *Lc1* locus and suggested that the *Lc1* locus consists of two loci, i.e., *qBF-A07-1* and *qBF-A07-2*. Further confirmation is required for this two-loci hypothesis. Hinchliffe et al. (2016) determined that the gene *GhTT2-A07* is the Lc1 gene that controls brown fiber in upland cotton. Recently, Yan et al. (2018) transformed this gene into cotton and confirmed its function of regulating fiber color formation.

Unlike brown cotton fiber, genetic control of green cotton fiber trait is complex and not well understood. It was proposed that the green cotton fiber is a quantitative trait. Previous genetic studies showed that the green fiber trait is dominant to the brown fiber trait, while the latter is dominant to the white fiber trait (Kohel and Richmond 1971; Kohel 1985; Shi et al. 2002). Wang et al. (2012) reported that green fiber trait is controlled by one major gene with incomplete dominance.

The molecular markers have been used to construct genetic maps, map important trait loci, characterize genetic diversity, and as well as determine seed purity. Cotton researchers used DNA markers to characterize color cotton germplasm in addition to map fiber color loci as mentioned above. As early as in 2001, Zhang and Wang (2001) used RFLP markers to characterize the differences between color cotton and white cotton. They speculated that genes controlling fiber color might also play roles in determining the quality property differences. Similar studies were conducted using random amplified polymorphic DNA, sequence-related amplified polymorphism, or simple sequence repeat markers (Guo et al. 2003; Ling 2009; Ma et al. 2003; Sun et al. 2009; Wang et al. 2012: You et al. 2014; Zhang et al. 2003; Zhang and Guo 2004). All demonstrated that a color cotton germplasm could be easily be differentiated from white cottons by using DNA markers. Diversities exist among color cotton germplasm from different regions, while differences within the same region are small. Wang and Li (2002) used amplified fragment length polymorphism (AFLP) markers to distinguish the parents used to make hybrid seeds of color cotton. These AFLP markers could be used to determine the hybrid seed purity.

6.4 Molecular Mechanisms of Fiber Color Formation

Same as white fibers, color cotton fibers emerge from the ovule epidermal cells. Its development includes four stages, i.e., initiation, elongation, secondary cell wall (SCW) biosynthesis, and maturation (see Chap. 7). In general, color cotton fibers have shorter elongation period and slower elongation speed than white fibers. Color cotton fiber development enters into SCW biosynthesis stage around 21 days post anthesis (DPA), while white fibers at around 16 DPA. The overlapping period between elongation and SCW is 5–6 days shorter in color fibers than the white fibers. The relatively inferior fiber quality of color cotton is likely related to this shortened duration. At around 20 DPA, large amount of cellulose is synthesized and accumulated in white fibers. However, at the same time in color fibers, pigments are synthesized, while less amount of cellulose is accumulated. In both color and white fibers, the fiber structure consists of primary cell wall, secondary cell wall, and lumen. However, there are big differences in cell wall thickness between color and white fibers. The thicknesses of primary cell wall, secondary cell wall, and lumen of white fibers are about 0.2 μm, 11.0 μm, and 6.5 μm, respectively. In brown fibers, they are 0.1 μm, 3.7 μm, and 10.3 μm, respectively. There are also some differences among different color fibers. The lumen of white fiber is usually hollow and barely contains anything, while pigments are present in the lumen of color fibers.

The chemical compositions of color fibers are very complex. Noncellulosic chemicals include wax, proteins, pectins, pigments, minerals, etc. It has been difficult to obtain an accurate characterization of the chemical components of color fibers, and reports from previous studies are conflicting. Ryser and Holloway (1985) identified tannin and its derivatives, as well as tannin's precursor catechin using chromatographic analysis. They suggested that the pigment in color cotton fiber was tannin derived from oxidization of catechin, which was supported by Chen et al. (2012). Yu (2002) suggested anthocyanins as the pigment in color cotton fibers. Zhao and Wang (2005) speculated that the pigment in brown cotton was a flavonoid but possible was a flavonol in green cotton. This opinion was supported by Tian et al. (2010), Ru (2010), and Li et al. (2012b). Zhan et al. (2004) extracted pigments in brown cotton fibers using 16 different reagents including acetone and conducted spectroscopic analysis. They found that the extracts had strongest and weakest absorbance at 218 nm and 264 nm, respectively. They concluded that the brown cotton pigment structure contains the double bond and phenol hydroxyl. Zhan et al. (2007) also conducted a quantitative investigation of the brown cotton fiber pigment content. The pigment in brown cotton was a quinone oxidized from a tannin. In summary, the color cotton fiber pigments are suggested as the following two types: (1) catechin oxidizes to tannin and then evolves to brown quinone and (2) the brown cotton fiber pigment is a flavonoid, while in green fiber it is a flavonol. It is worth to note that to determine the actual identities of color cotton fiber pigments requires not only chemical analysis of fibers but also the gene network related to pigment synthesis and formation. With good understanding of fiber color formation, it will be possible for us to manipulate fiber color while maintain good quality attributes.

Different cotton fiber colors are probably controlled by different genes. The variable intensity of the same color such as brownness might be due to incomplete dominance or additive effects (Liu et al. 2015; Zhang et al. 2011). Cotton fiber colors are mainly due to genetics with minor environment effects (Gong et al. 2014).

The biosynthetic pathway of anthocyanin in plants is a branch of flavonoid biosynthesis pathway. Many genes and enzymes such as chalcone synthase (CHS), chalcone isomerase (CHI), flavanone 3-hydroxylase (F3H), dihydroflavonol 4-reductase (DFR), flavonoid 3′-monooxygenase (F3'H), flavonoid 3'-5'-hydroxylase (F3'5'H), anthocyanidin synthase (ANS), anthocyanin reductase (ANR), and leucoanthocyanidin reductase (LAR) are involved in this pathway. A comparative analysis between a color cotton and its near isogenic white cotton line showed that the expression levels of *Gh3GT* (*Gh*Flavonoid-3-O-glucosyltransferas), *GhF3'5'H*, and *GhCHS3* genes were significantly higher in color cotton fibers than in white fibers. Xiao et al. (2007) examined the expressions of five genes (*CHI*, *F3H*, *DFR*, *ANS*, and *ANR*) in the flavonoid biosynthesis pathway in both T586 brown cotton and white cotton. They observed much higher expression of these five genes in 8, 16, 20, and 30 DPA brown fibers than in white fibers. Sun (2013) obtained similar results when studying *GhANS* gene expression in 15 DPA fibers of both color and white cotton fibers. As mentioned above, the actual chemical identity of brown cotton fiber pigment is still not clear. Anthocyanin might be one candidate. Tan et al. (2013) transferred the *F3H* gene into a brown cotton. RNAi suppression of F3H gene did reduce the amount of proanthocyanidins in the transgenic cotton fibers; however, the fibers were actually darker. Therefore whether brown cotton fiber pigment is proanthocyanidins or oxidized proanthocyanidins requires further investigation.

The biosynthesis of anthocyanin or proanthocyanidins requires not only structure genes but also many transcription factors. These transcription factors include three types, myeloblastosis (MYB) bHLH protein and WD40 protein. Yan et al. (2018) identified five pairs of cotton *TT2* homologous genes in the brown cotton T586 genome and found that all these five pairs of genes could promote the synthesis of proanthocyanidins. They also discovered that *GhTT2-3A* is the *Lc1* gene. The *Lc1* gene is known to control the formation of brown color in cotton fibers. Liu (2014) reported that there were 294 bHLH proteins in the *Gossypium* genome, and five of them were orthologous to *Arabidopsis* bHLH genes that were known to be involved in the biosynthesis of proanthocyanidins. One of them was *GhbHLH110A* whose ortholog was *TT8* gene (Yan 2016). Either *GhTT2-3A* or *GhbHLH110A* could activate the expression of *GhDFR*, *GhLAR*, or *GhANR* promoters, but the effects were quite weak. When combining *GhTT2-3A* and *GhbHLH110A*, the expression of these promotes was significantly enhanced.

6.5 Color Cotton Germplasm Improvement

The yield and fiber quality of color cottons lag behind white cottons, which impedes large-scale utilization of color cottons. The negative correlation between fiber colors and quality is one of the major difficulties that a cotton breeder encounters when using a regular cross-selection method to improve color cotton germplasm One breeding strategy that breeders tried and achieved success is to make interspecific cross to help break negative association between fiber color and length (Feng 2006). In China, breeders crossed semi-wild landlaces or racestocks in combination of backcrossing. In 1987, Du crossed *G. hirsutum race richmondi*, a semi-wild species native to Mexico, with white upland cotton Texas Marker-1 (TM-1). After several years of backcrossing, selection, and acclimatization, he found a brown cotton plant that was similar to wild parent in architecture but to TM-1 in agronomic performance. This plant also had shorter growing season than either parents. After rigorous field tests, this new brown cotton line "Brown 1" was registered and released. In addition, Chinese breeders artificially treated seeds using radiation and obtained fiber mutations. Then they crossed the mutants with normal white cotton, and bred new brown cotton line "Brown 263" (Fig. 6.2) (Du et al. 2002). This variety is characterized for its stable fiber color and high yield, but its quality is inferior to "Brown 1." Another line "Brown 3-994" has high gin turnout, comparable yield, and fiber quality as white cottons. Feng (2003) made a series of crosses using color cottons (brown or green) as female parents and white sea-island cotton as the male parent. He has identified F_1 plants that have optimal fiber colors and good fiber length and other quality attributes.

Increasing yield through hybrid heterosis has been widely used in cotton production especially in China and India. The same strategy was also applied in color cotton production. The major hindrance for utilizing hybrids is the cost associated with F_1 seed production. At present, there are three ways to produce F_1 seeds in mass. The first is "artificial emasculation and pollination" which is widely used in India. This is a very expensive method and can only be implemented when labor cost is low. The second way is to use nuclear male-sterile line that is used in Sichuan province, China. The drawback of this method is the instability of male-sterile line. The third type is to use cytoplasmic male-sterile line. In the 1970s, US scientists discovered a cytoplasmic male-sterile line in *G. harknessii*. But its restoration line had weak ability to restore the fertility in F_1 plants, resulting in very low rate of boll setting and poor yield. Thus it did not attract much attention in breeding and production until early 2000. Wang and Li (2002) transferred the glutathione S transferase gene into a restoration cotton line. After screening transgenic plants, a cotton line with strong restoration ability was selected. Breeding of this restoration line made utilization of cytoplasmic male-sterile line in F_1 seed production possible in commercial scale. Currently, cotton researchers in China are using this method to make hybrids between color cotton and sea-island cotton. Whether a hybrid color cotton can achieve success in the market place mainly depends on its gin turnout. Gin turnout is a quantitative trait and controlled by many genes. Selection of parents is critical when making an F_1 hybrid. A general rule is to choose high gin turnout color cotton line as a female parent and a white cotton with early maturity and high gin turnout as male parent. Dutt et al. (2004) made a cross between a brown cotton male-sterile line and a white cotton restoration line. The F_1 plants achieved good yield and quality besides the intended color. Scientists in the Chinese Institute of Cotton Research used the Bt transgenic cotton variety 971201 (a selection from ZM41) as female and brown cotton RILB263102 as male parent. The hybrid variety, ZM51, was characterized as high yield, pest insect resistance, and good quality. Its yield was 1095.0 kg/ha, and quality attributes were average length 30.5 mm long, strength 30.4 cN/tex, MIC 4.5, uniformity 84.7%, and elongation 6.7% (Li et al. 2009).

Improving color cotton fiber quality using biotechnologies has attracted great attention. Scientists in Chinese Academy of Science created blue and red cotton fibers by modifying target genes. They cloned the gene *indole dioxygenase* (*IDG*) responsible for blue fiber pigment formation and gene *dihydroflavonol 4-reductase* (*DFR*) regulating red anthocyanin synthase. They made a construct by joining these two genes with fiber-specific promoter PTL12 and transformed the construct into a cotton plant. The transgenic cotton plants produced cotton bolls with blue or red fibers. This research opened a new avenue to obtain new fiber colors while maintaining fiber quality and yield potential.

Pest resistance is important in modern cotton production. Many white cotton varieties contain a Bt gene that confers resistance to chewing insects such as bollworms. In order for a color cotton variety to be planted in large acreage, to have a resistance to prevailing pests is important. Toward this goal, Chinese breeders crossed a Bt-containing white cotton variety 971201 with brown cotton line RILB263102 and bred a color cotton hybrid, ZM51. This hybrid is resistant to chewing insects and also tolerant to verticillium wilt and fusarium wilt (Sun et al. 2005). The researchers in Sichuan Academy of Agricultural Sciences introgressed the Bt gene into a brown cotton and bred the variety “Chuan Cai 3” (Gong et al. 2008). Simultaneously, they bred a Bt-containing green cotton variety “Chuan Cai 4.” Both varieties are tolerant to verticillium wilt and fusarium wilt besides containing a Bt gene (Ji et al. 2009). Zhang et al. (2005) transformed a three-gene construct Bt-Cpt1-GNA into a color cotton plant and obtained transgenic color cotton plants that are resistant to cotton bollworms.

Besides the ICR-CAAS, several organizations in China have been conducting color cotton research and varietal improvement. Jiangsu Academy of Agricultural Science released a high-quality hybrid color cotton variety “Su Caiza 1.” Its average fiber length was 29.8 mm, strength 30.4 cN/tex, and MIC value 4.3. The variety “Xincai 25” was bred and released by Chinese Color Cotton LLC. This variety has brown fiber length 29.0 mm, strength 29.28 cN/tex, and MIC value 4.12. Guo et al. (2008) crossed five upland color cotton lines with six white cotton lines having superior fiber properties. They obtained several high-yield F_1s with good fiber quality. A few of them are being used for commercial production.

Compared to brown cotton, the green color of cotton fiber is less stable. Varietal improvement of green cotton has been difficult. In China, researchers at the ICR-CAAS released a green cotton variety “Green 1” in 1993 through pedigree selection from an American green cotton line. This variety had short, weak, and fine green fiber. Zhang et al. (2007) began crossing green cotton line Lnccx-7 with white cotton 1-713-1.26 and obtained a green cotton variety “Yun Cai Lv 1.” This variety had higher yield and better quality than “Green 1.” In addition, it had resistance to bollworms and verticillium wilt. The Gansu Institute of Agricultural Sciences bred three green cotton varieties “Longlv 1,” “Longlv 2,” and “Longlv 3” that are suitable for local climate. Green cotton variety “Su Lv 137” bred by Chen et al. (2015) has fibers with mean length 29.4 mm, uniformity 84.7%, strength 27.1 cN/tex, elongation 6.3%, and MIC value 2.8. The “Xincai 16” bred by Chinese Color Cotton LLC and Xinjiang Institute of Agricultural Sciences is a high-yield green cotton variety in Xinjiang province. Its yield could reach 1491.3 kg/ha. Its fiber has mean length 30.65 mm, uniformity index 82.99, strength 24.85 cN/tex, elongation 6.47%, and MIC value 2.75.

6.6 Color Cotton Production and Utilization

The planting and management techniques for color cotton are similar to those for white cotton. One important thing to keep in mind is isolation when planting color cottons. This includes isolation from white cotton and other color cottons having different colors. Isolation planting is important in order to avoid cross contamination. Isolation can be achieved through either physical isolation such as using a covering net or planting in distance. The former is only feasible for small experimental tests. The latter is commonly used in commercial production. The field for seed production should be far away enough from any other cotton fields to avoid contamination by other cotton pollens. The planting density varies from region to region. In Xinjiang and other Northern China regions, the planting density is usually about 150,000 plants per hectare, while in Yellow River region, it is about 60,000 per hectare.

Compared to white cotton, the advantage of using color cotton is due to its low carbon footprint to the environment. Thus, it will be imperative to keep any forms of pollution to the minimum when producing and utilizing color cotton fibers. This starts from planting a variety that has built-in resistance to pests including insects, weeds and diseases. Second is to develop and maintain a good growing environment that an integrative pest management system is able to be implemented. Whenever possible, pest control practices should be relying on natural and biological means first. Applying pesticides should be minimum and low toxicity to humans and animals. For fertilization, use as much organic fertilizers as possible. Third is to optimize the growing practices. For example, inter-planting other crops may be helpful to control weeds. It should be noted that the practices should be optimized based on the local situations.

Although cotton fiber colors are mainly determined by genetics, environment factors such as sunlight and soil types also affect fiber color especially color intensity. It is very common that fibers of the same variety but growing in different environments have different color intensities. It is not unusual to observe fiber color segregation among plants of the same field and sometimes between bolls of the same plant. In general, the quality of light color fibers is comparable to white fibers, while dark color fibers are usually inferior in quality. Sulfur-containing atmosphere or acid rain greatly affects fiber color development. A green cotton variety growing under a high-sulfur environment may produce brown fibers. Thus, application of a sulfur-containing fertilizer or pesticide should be limited during fiber development stage. Fiber color may change if high moisture occurs during harvest season. An inappropriate storage condition may also alter fiber colors. The naturally colored cotton fibers require different processing methods and chemical agents in order to protect the integrity of the nature color. Fiber colors usually become darker after regular textile processes such as boiling, acidic treatment. Bleach or other solutions containing oxidants can make fiber color fade or disappear completely.

In order to improve gloss and feel of cotton fibers, it is a common practice to treat cotton fibers with an alkaline solution (see Chap. 5). For color fibers, certain amount of pectolase and cellulase may be added to remove the residue pectin (Zhu et al. 2006). This process does reduce fiber or fabric color intensity as colorless alkaline solution becomes colorful after treatment. In general, after treatment a green cotton fabric maintains its nature color better than a brown cotton fabric.

As mentioned above, sunlight illumination greatly alters natural colors of cotton fibers or fabrics. Green color will gradually change to yellowish green then to yellowish brown even under indoor nature light condition. In brown cotton, it is often to see color difference between the fibers exposed to and not exposed to sunlight. The former is light brown while the latter reddish brown. Li and Zhu (2003) compared the fabric color stability between nature fiber colors and artificially dyed colors. They showed that artificially dyed fabrics lost colors quicker and more severe than fabrics made from naturally colored cotton fibers.

There are differences in physical and chemical properties between color and white fibers. Color fibers usually contain higher amount of wax, thus has lower absorbance than white fibers. The moisture maintaining abilities of white, brown, and green fibers are 8.5%, 7.6%, and 5.1%, respectively. Although color cotton fiber quality has improved significantly during the past two to three decades, a better quality is required to meet the demands of ever-evolving textile-spinning technologies.

Natural color cotton fibers are mainly used in underwear, night clothes, children's clothing, and jeans. Color fibers are frequently blended with white fibers and sometimes with synthetic fibers. A wide range of colors can be achieved by adjusting the blending ratios of color fibers and white fibers.

References

Chen XS, Di JC, Zhao L (2015) Breeding and product development of high quality colored cotton. Cotton Textile Technol 43:81–84

Chen YC, Wu T, Liu HF et al (2012) Extraction and Identification of proanthocyanidins from brown cotton fiber. Acta Agric Boreali-Occiden Sin 21:92–96

Deng FJ, Chen GW, Yu Y et al (2010) The technological progress of the cotton industry over the past 30 years in Xinjiang Production and Construction Corps. Xinjiang Farm Res Sci Technol 6:3–7

Du XM, Sun JL, Zhou ZL et al (2002) The variety improvement and research progress on fiber of naturally color cotton. In: 2002 Annual Meeting of China Cotton Society, pp 110–113

Dutt Y, Wang XD, Zhu YG, Li YY (2004) Breeding for high yield and fiber quality in coloured cotton. Plant Breed 123:145–151

Endrizzi J, Taylor T (1968) Cytogenetic studies of N Lc1 yg2 R2 marker genes and chromosome deficiencies in cotton. Genet Res 12:295–304

Feng HJ, Wang J, Sun JL et al (2010) Genetic effects of fiber color in brown cotton (*Gossypium hirsutum* L.). Acta Agron Sin 36:961–967

Feng K (2003) The segregation of fiber traits of distant hybrid progeny of color cotton. Gansu Agric Sci Technol 10:19–20

Feng KY (2006) A study on distant hybridization of *Gossypium hirsutum* L. and *Gossypium barbadense* L. Chin Agric Sci Bull 12:181–183

Geng JY, Wang GY, Zhai XJ et al (1998) Effects of colored fiber gene on economic properties of upland cotton and analysis of its inheritance. Cotton Sci 6:307–311

Gong WF, He SP, Tian JH et al (2014) Comparison of the transcriptome between two cotton lines of different fiber color and quality. PLoS One 9:e112966. https://doi.org/10.1371/journal.pone.0112966

Gong YY, Ji PC, Yang ZH, et al (2008) Selection of Chuan color cotton NO.3 with insect-resistant. In: 2008 Annual Meeting of China Cotton Society, pp 240–242

Guo BD, Huang SL, LX J et al (2008) Analysis of combing ability on yield and quality traits in nature color cotton. J Shanxi Agric Sci 36:11–13

Guo JY, Wang YQ, Wu MG (2003) Genetic diversity analysis and identification of colored cotton germplasms by RAPD markers. Cotton Sci 15:269–273

Hinchliffe DJ, Condon BD, Thyssen G et al (2016) The *GhTT2_A07* gene is linked to the brown colour and natural flame retardancy phenotypes of *Lc1* cotton (*Gossypium hirsutum* L.) fibres. J Exp Bot 67:5461–5471

Hu BT (2004) Status quo and prospect of natural colored cotton. Cotton Textile Technology 32:27–31
Ji PC, Mou FS, Mao ZX, et al (2009) Selection of Chuan color cotton NO.4 with insect-resistant. In: 2009 Annual Meeting of China Cotton Society, pp 188–190
Kohel RJ (1985) Genetic analysis of fiber color variants in cotton. Crop Sci 25:793–797
Kohel RJ, Richmond TR (1971) Isolines in cotton: effects of nine dominant genes. Crop Sci 11: 287–289
Li DG, Nie YC, Zhang XL (2004) Genetic analysis of fiber color on brown upland cotton. J Huazhong Agric Univ 23:606–609
Li FZ, Ning XM, Qiu XM et al (2012a) Genetic mapping of the dark brown fiber lc1 gene in tetraploid cotton. Sci Agric Sin 45:4109–4114
Li HX, Tang ZR, Li D et al (2012b) Extraction of pigments in naturally green cotton and spectral properties research. J Zhejiang Sci-Tech Univ 29:32–35
Li JL, Ning XZ, Lin H (2009) Analysis on heterosis of color cotton - conventional cotton hybrid F1. Xinjiang Farm Res Sci Technol 4:8–10
Li JQ, Zhu MH (2003) Study on the color stability and fastness to weathering of naturally color cotton. China Fiber Inspection 12:22–24
Ling L (2009) Genetic difference analysis between colored cotton and white cotton by SRAP markers and morphological markers. Dissertation, Anhui Agricultural University
Liu HS (2014) Cloning, expression and bioinformatic analysis of bHLH protein genes in cotton. Dissertation, Southwest University
Liu SJ, Gong WF, Sun JL et al (2015) Effects of growth substances on the fiber growth and development of color cotton ovule culture *in vitro*. Sci Agric Sin 48:2127–2142
Ma X, Du XM, Sun JL (2003) SSR fingerprinting analysis on 18 colored cotton lines. J Plant Genet Resour 4:305–310
Pan ZE, Du XM, Sun JL et al (2006) Influences of boll shading on fiber color and fiber quality of colored cotton. Cotton Sci 18:264–268
Ru ZL (2010) Research on the ultra-microstructure and chemical properties of natural color cotton. Dissertation, Chinese Academy of Agricultural Sciences
Ryser U, Holloway PJ (1985) Ultrastructure and chemistry of soluble and polymeric lipids in cell walls from seed coats and fibers of Gossypium species. Planta 163:151–163
Shi YZ, Du XM, Liu GQ et al (2002) Genetic analysis of naturally colored lint and fuzz of cotton. Cotton Sci 14:242–248
Sun DL, Sun JL, Du XM et al (2008) Genetic study on the color of fiber and linter in brown cotton. J Anhui Agric Sci 36:6254–6255
Sun DL, Sun JL, Du XM et al (2009) Comparative study of genetic diversity in brown and green cotton. Acta Agri Boreali Sin 24:96–101
Sun HW (2013) Gene cloning and functional analysis of colored cotton flavonoids metabolism-related genes *GhANS and GhCHS*. Dissertation, Shihezi University
Sun JL, Zhou ZL, Pan ZE et al (2005) National authorized color cotton variety--ZM51. China Cotton 11:22–23
Tan J, Tu L, Deng F et al (2013) A genetic and metabolic analysis revealed that cotton fiber cell development was retarded by flavonoid naringenin. Plant Physiol 162:86–95
Tan YL, Zhou XZ (2015) The research status and development trends of natural colored cotton. Prog Text Sci Technol 2:1
Tian XH, Li YJ, Zhang XY et al (2010) HPLC determination of myricetin, quercetin, naringenin, and kaempferol in the natural brown cotton fiber. Cotton Sci 22:597–602
Wang L, Liu H, Li X et al (2014) Genetic mapping of fiber color genes on two brown cotton cultivars in Xinjiang. SpringerPlus 3:1–5
Wang LX, Liu HF, Xiao XW et al (2012) Genetic properties of colored cotton in Xinjiang. J Anhui Agric Sci 13:3926–3929

Wang XD, Li YY (2002) Development of cytoplasmic male-sterile, maintainer and restorer lines in colored cotton and construction of their DNA fingerprints. J Zhejiang Univ (Agric Life Sci) 28:1–6

Ware JO (1932) Inheritance of lint colors in upland cotton. J Am Soc Agron 24:550–562

Wen TW, Wu M, Shen C et al (2018) Linkage and association mapping reveals the genetic basis of brown fibre (Gossypium hirsutum). Plant Biotechnol J. https://doi.org/10.1111/pbi.12902

Xiao YH, Zhang ZS, Yin MH et al (2007) Cotton flavonoid structural genes related to the pigmentation in brown fibers. Biochem Biophys Res Commun 358:73–78

Yan Q (2016) Map-based cloning of dark brown fiber gene *Lc1*, and the regulation of proanthocyanidin biosynthesis and brown fiber coloration. Dissertation, Southwest University.

Yan Q, Wang Y, Li Q et al (2018) Upregulation of *GhTT2-3A* in cotton fibers during secondary wall thickening results in brown fibers with improved quality. Plant Biotechnol J. https://doi.org/10.1111/pbi.12910

You CY, Nie XH, Zhang S et al (2014) Construction of DNA fingerprinting and analysis of genetic diversity with SSR markers for 23 color cotton cultivars in Xinjiang. Cotton Sci 26:161–170

Yu B (2002) Color fixation of naturally green cotton with metallic salts. Knitting Ind 30(1):35–37

Zhan SH, Li ZP, Lin Y et al (2008) Quantitative analysis on the inherited characteristics of naturally colored brown cotton fiber color. Chin Agric Sci Bull 24:146–148

Zhan SH, Lin Y, Cai YP et al (2004) Extraction and purification of the pigment of natural brown cotton and its UV spectral properties. Acta Laser Biol Sin 13:324–328

Zhan SH, Lin Y, Cai YP et al (2007) Preliminary deductions of the chemical structure of the pigment brown in cotton fiber. Chin Bull Bot 24:99–104

Zhang BX, Wang GX (2001) The differences of DNA molecular markers between color cotton and white cotton. China Cotton 28:16–17

Zhang LS, Wu X, Shangguan XX et al (2007) A new color cotton variety- Yuncailv-2. China Cotton 34(12):18–19

Zhang LS, Zhu Z, Wu X et al (2005) Acquisition of transgenic colour cotton with a trivalent insect-resistant gene. Acta Botan Boreali-Occiden Sin 25:1126–1131

Zhang MD, Zhan XJ, Zhang XL (2003) Polymorphic analysis on natural color fiber cotton by RAPD. J Huazhong Agric Univ 22:427–430

Zhang ML, Song XL, Sun XZ et al (2011) Observation of differentiation and pigment deposition process in colored cotton fibers. Acta Agron Sin 37:1280–1288

Zhang QY, Guo JY (2004) Breed identification of colored cotton germplasm lines by RAPD markers. Xinjiang Agric Sci 41:143–146

Zhang XL (1998) Preliminary report of color cotton breeding. China Cotton 25:18

Zhang XZ, Zhou YS, Lan JX (2002) Current status of research and genetic analysis of some traits of naturally colored cotton. Chin Agric Sci Bull 18:92–93

Zhang ZS, Hu MC, Zhang J et al (2009) Construction of a comprehensive PCR-based marker linkage map and QTL mapping for fiber quality traits in upland cotton (*Gossypium hirsutum* L.). Mol Breed 24:49–61

Zhao XQ, Wang XD (2005) Composition analysis of pigment in colored cotton fiber. Acta Agron Sin 31:456–462

Zhu XY, Li J, Zheng T (2006) Study on mercerizing treatment of colored cotton knitted fabric. Shandong Text Sci Technol 5:1–4

Chapter 7
Cotton Fiber Biosynthesis

Hee Jin Kim

7.1 Introduction

Cotton is the most important agricultural textile commodity in the world (Wakelyn et al. 2010). Cotton fibers are unicellular trichome cells arising from the outer integument layer of the seed coat. Due to the highly elongated primary cell wall (PCW) structure and secondary cell wall (SCW) containing almost pure cellulose, cotton fibers have been used to study cell differentiation, cell elongation, and cell wall biogenesis (Haigler et al. 2012; Kim and Triplett 2001). Among more than 50 recognized cotton species consisting of diploid and allotetetraploid (Wendel et al. 2012), *Gossypium hirsutum* L. (AD_1 genome), known as upland cotton producing high-yield and generally good fiber properties, is growing in greater than 90% of the world's cotton production (Zhang et al. 2008). The remaining cotton supply is produced from the closely related *G. barbadense* (AD_2 genome), *G. herbaceum* (A_1 genome), and *G. arboreum* (A_2 genome) species. Draft genome sequences from the cultivated *G. arboreum* (Li et al. 2014), *G. hirsutum* (Li et al. 2015; Zhang et al. 2015), and *G. barbadense* (Liu et al. 2015; Yuan et al. 2015) as well as a reference genome (D_5) sequence from non-spinnable *G. raimondii* (Paterson et al. 2012) have been reported. Along with the cotton genome sequences, numerous transcriptome profiles have expanded our understanding of molecular mechanisms regulating cotton fiber development, and they have been reviewed at the Chaps. 8 and 9 in this book as well as other reviews (Alleman and Allen 2010; Guan and Chen 2013; Kim 2015; Stiff and Haigler 2012; Stiff et al. 2016; Udall 2009).

This review chapter covers the recent advances in understanding of cotton fiber development at the four different developmental stages using both conventional and

H. J. Kim (✉)
Cotton Fiber Bioscience Research Unit, USDA-ARS, Southern Regional Research Center, New Orleans, LA, USA
e-mail: HeeJin.Kim@ars.usda.gov

D. D. Fang (ed.), *Cotton Fiber: Physics, Chemistry and Biology*,
https://doi.org/10.1007/978-3-030-00871-0_7

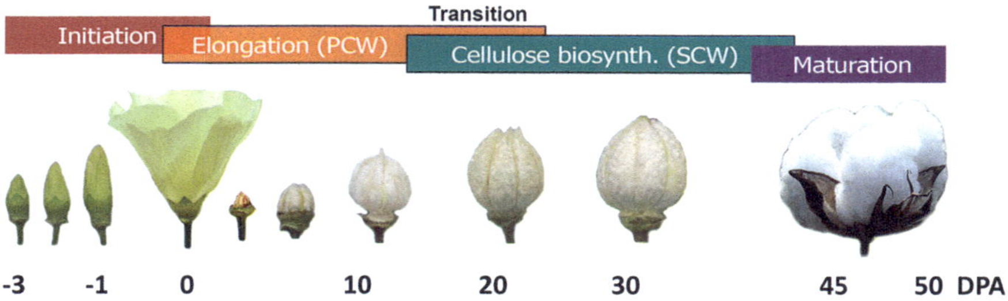

Fig. 7.1 *G. hirsutum* boll and fiber development. Upland TM-1 cotton fiber developmental stages are classified according to physical properties including fiber length, cellulose content, cell wall thickness, chemical components, crystallinity, and cellulose microfibril assembly of developing cotton fibers grown in a cotton field located in New Orleans, Louisiana. The figure was modified from Liu and Kim (2017). *PCW* primary cell wall, *SCW* secondary cell wall, *DPA* days postanthesis

advanced phenotypic methods. It also describes how the fiber development affects the physical properties and quality of cotton fibers with minimum redundancies with other chapters. An upland cotton variety, Texas Marker-1 (TM-1), was mostly used in this chapter because it is a genetic standard upland variety for cotton breeders and geneticists (Kohel et al. 1970; Zhang et al. 2015).

Cotton fibers grow inside of a developing cotton boll (Fig. 7.1). Cotton fiber development is classified into four overlapping stages: (1) initiation, (2) PCW biosynthesis for fiber elongation, (3) SCW biosynthesis for cellulose production and cell wall thickening, and (4) maturation and cell death process before the boll opens (Naithani et al. 1982).

Cotton lint fiber initials differentiate from −1 or 0 days postanthesis (DPA) ovules from *G. hirsutum* TM-1. Around 3–5 DPA, another class of fiber cells, fuzz fibers, or linters starts growing (Lang 1938). A large central vacuole becomes prominent at early stage of cotton fiber development. During formation of the PCW, a thin PCW is deposited in elongating fibers. Fiber elongation continues to approximately 21–26 DPA. The lint grows up to 2.2–3.6 cm, whereas the linter grows up to 1.2–1.5 cm. At approximately 14–16 DPA, fiber cells synchronously enter the SCW stage overlapping the final PCW stage. During the transition from PCW to SCW biosynthesis, synthesis of other cell wall polymers ceases, and the rate of cellulose synthesis in cotton fibers dramatically increases (Meinert and Delmer 1977). Mature fibers exhibit thickened SCW and consist of 90–95% cellulose. General characterizations of cotton fiber properties and anatomy during cotton fiber development can be found in other reviews (Benedict et al. 1999; Haigler 2010; Haigler et al. 2012; Kim and Triplett 2001; Meinert and Delmer 1977; Ryser 1999).

To interpret the transcriptomic results that were obtained from RNA extracted from developing fibers at early developmental stage, cotton researchers need accurate and quantitative phenotype data (length, strength, maturity, fineness, cellulose properties) from the fiber samples that they used for transcriptomic analyses. However, the physical properties of developing fibers that are immature and sticky due to high contents of physiological sugars and pectins cannot be measured by high-volume instrument (HVI) and Advanced Fiber Information System (AFIS) that are designed to measure fully developed fibers and have practical limits in use for biological research (Kim et al. 2014; Paudel et al. 2013). In addition, a few other fiber-specific methods including Stelometer (fiber strength), Cottonscope (fiber maturity and fineness), gravimetric fineness (fineness), and Favimat (single fiber breaking force and strength) have been used (reviewed in Chap. 3). To understand physical properties of fiber walls and estimate the amount of cellulose amount, cellulose content, and crystallinity as well as microfibril orientation, general laboratory methods including a wet chemical assay for cellulose (Updegraff 1969), cross-sectional image analysis microscopy (IAM), polarized optical microscopy of longitudinal fiber arrays (Chap. 3), and analytical methods including Fourier transform infrared (Chaps. 4 and 5) and X-ray powder diffraction (Chaps. 2 and 5) have been used. Recently, several new methods, including Raman, and sum-frequency generation (SFG) spectroscopic methods (Kim et al. 2017; Lee et al. 2015) and glycome profiling (Avci et al. 2013) are also used to determine chemical components, cellulose microfibril packing, and carbohydrate levels from developing fibers at different developmental stages. These advanced phenotypic techniques as well as the fiber-specific instruments can greatly improve the ways of determining accurate phenotypes of cotton fibers, interpreting transcriptomic data, and identifying the molecular mechanisms of SCW cellulose depositions that affect cotton fiber properties and quality.

7.2 Differentiation of Fiber Initials

Upland lint fiber initials are differentiated at the chalazal end of the ovules from the ovular epidermal cells during the fiber initiation stage (Fig. 7.2); new fiber initials are progressively differentiated toward the micropylar end (Lang 1938; Ryser 1999; Stewart 1975). The number of cotton ovule epidermal cells differentiating into fiber

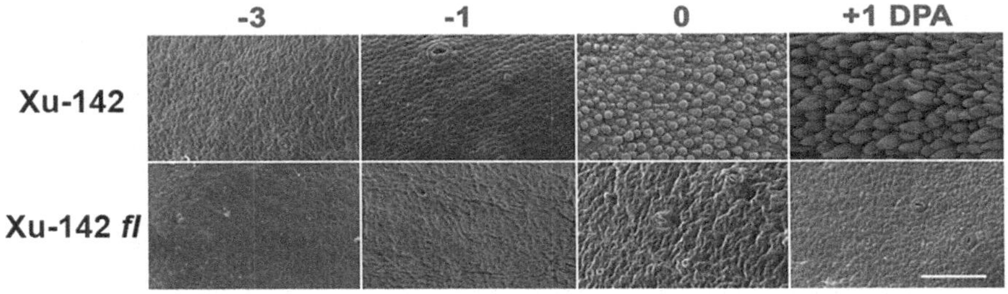

Fig. 7.2 Differentiations of fiber initials. Scanning microscopic images of epidermal tissue were taken from *G. hirsutum* Xu-142 ovules with fiber initials (top panel) and fiberless mutant, Xu-142 *fl* (bottom panel). The ovules were harvested around 9 AM on −3, −1, 0, and + 1 DPA. The bar represents lengths of 50 μm. The figures were modified from Kim et al. (2015)

initials was suggested as an indicator of cotton yield (Li et al. 2009). A higher density of fiber initials presumably caused by higher auxin levels in the ovule epidermis at the fiber initiation stage resulted in higher fiber yield and finer fibers (Zhang et al. 2011). Thus, the fiber initiation stage has become a potential target for improving yield by biotechnological manipulations (Lee et al. 2007).

Fiber initial differentiation process is quasi-synchronized in each developing ovule and among ovules, and cotton ovules between 0 and 4 DPA contain a mixture of fiber initials and elongating fibers (Basra and Malik 1984). Approximately 15–30% of epidermal cells on cotton ovules are differentiated to fiber initials (Applequist et al. 2001; Lee et al. 2007; Stewart 1975). Around 5–10 DPA, fuzz fibers or linters are differentiated (Lang 1938).

Due to the potential for improving cotton yield and quality by increasing the number and/or density of fiber initials, multiple groups searched for cotton genes regulating the differentiation of fiber initials from the epidermis of cotton ovules using natural fiberless mutants or fuzzless and lintless (*fl*) mutants like Xu-142 *fl* (Zhang and Pan 1991) and MD17 and SL 1–7-1 (Turley and Kloth 2002; Turley and Kloth 2008) as well as fuzzless mutants such as N_l (Kearney and Harrison 1927; Lee et al. 2006) and n_2 (Ware et al. 1947). The n_2 gene and li_3 gene are related to the lintless and fuzzless phenotype of Xu-142 *fl* mutant (Ma et al. 2016; Wu et al. 2007, 2006).

GhMYB25-like transcription factor (Walford et al. 2011), a GhHD-1 transcription factor (Walford et al. 2012), and a vacuolar invertase, GhVIN1 (Wang et al. 2014) are involved in differentiation of fiber initials. Later, a map-based cloning identified the *GhMML3_A12* encoding *GhMYB25-like transcription factor* as the N_l that is associated with fuzz fiber development (Wan et al. 2016) and *GhMML4_D12* encoding another *GhMYB transcription factor* as Li_3 that is associated with lint fiber development (Wu et al. 2017).

In addition, reactive oxygen species (Zhang et al. 2010) and small RNAs (Pang et al. 2009; Wan et al. 2016; Xie et al. 2015) have been suggested to play important roles in the fiber initiation process. Phytohormonal signaling networks regulate the temporal expression of genes responsible for differentiation of cotton fiber initials (Kim et al. 2015; Yang et al. 2006). Auxin and gibberellic acid are essential for differentiating fiber initials in vitro (Beasley and Ting 1974). A jasmonic acid signaling protein negatively regulates fiber initiation by interacting with GhMYB25-like (Hu et al. 2016).

7.3 Fiber Elongation in *G. hirsutum*

Cotton fiber elongation process has been extensively characterized due to relatively easy ways of measuring the lengths and extracting high quality of transcripts from developing fibers at the elongating stage as well as the importance of fiber length determining quality and commercial values of raw fibers and yarns.

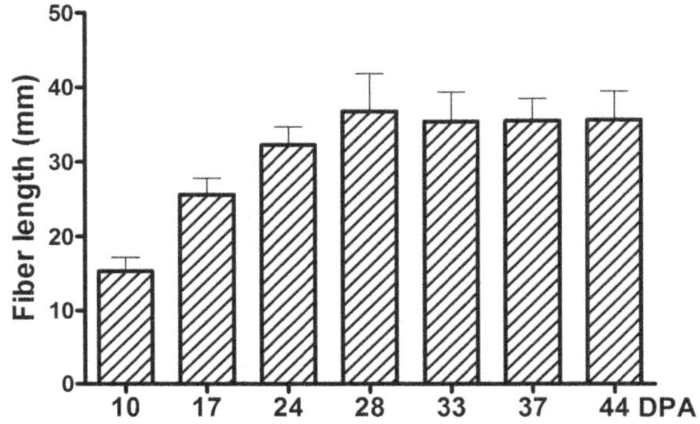

Fig. 7.3 Fiber elongation and cellulose biosynthesis during fiber development. Average lengths of developing fibers at various DPAs from *G. hirsutum* TM-1 were measured. The figures were modified from Kim et al. (2013b)

The differentiated fiber initials elongate rapidly during active PCW stage (0–17 DPA) (Fig. 7.3). In elongating fibers, a thin PCW is deposited. The Raman spectra (Fig. 7.4a) of developing TM-1 fibers show that the 10 DPA fibers contain long-chain alkyl components of the cuticle and pectin (2883 and 2933 cm^{-1}) and unsaturated hydrocarbon components (1614 cm^{-1} and 1660 cm^{-1}) (Kim et al. 2017). The negative principal component (PC1) scores indicated that 10 DPA cotton fibers were mostly PCW (Fig. 7.4a). Fiber PCW is a composite of cellulose, hemicellulose, pectin, and structural wall proteins. The cellulose content of the expanding PCW is less than 10–15% by weight. As the transition stage from PCW to SCW stage begins at approximately 14–16 DPA, the fiber elongation rate slows down although the developing TM-1 fibers continue elongating until 24–28 DPA (Fig. 7.3).

Recent observations support a notion that fiber cells may expand via a linear growth mode, which is the combination of the diffuse growth modes and tip growth (Qin and Zhu 2011). Cotton fibers elongate as tissue-like bundles, not as individual cells (Singh et al. 2009). At the end of the fiber elongation stage, cotton fibers become separated from one another and elicit high enzyme activities related to cell wall hydrolysis. During fiber development, microtubules exhibit specific changes in orientation, organization, number, length, and proximity to the plasmalemma. In expanding cotton fibers, the patterns of microtubule deposition correlate precisely with the wall microfibril arrays (Seagull 1986, 1993). The rapid fiber elongation is driven by high turgor pressure that is regulated by developmentally reversible gating of plasmodesmata and a coordinated activity of solute transporters and wall-loosening expansins (An et al. 2007; Ruan et al. 2001, 2004; Shi et al. 2006).

Functional analyses of multiple genes including *GhACT1* involved in actin cytoskeleton network (Li et al. 2005), annexin (Zhang et al. 2016), *GhMYB25* (Machado et al. 2009), *GhMYB109* (Pu et al. 2008), *PDF1* (Deng et al. 2011), $GhGA20_{ox}1$ (Xiao et al. 2010), and sucrose synthase (Ruan et al. 2003) showed their involvements in fiber elongation process. Later, mapping-by-sequencing identified a single amino acid mutation of an actin gene (Gh_D04G0865) causing the short fiber phenotype of the *Ligon-lintless 1* (Li_1) mutant (Thyssen et al. 2017).

In addition, phytohormonal regulations by ethylene (Shi et al. 2006), gibberellin (Aleman et al. 2008; Liao et al. 2009), and brassinosteroids (BR) (Luo et al. 2007;

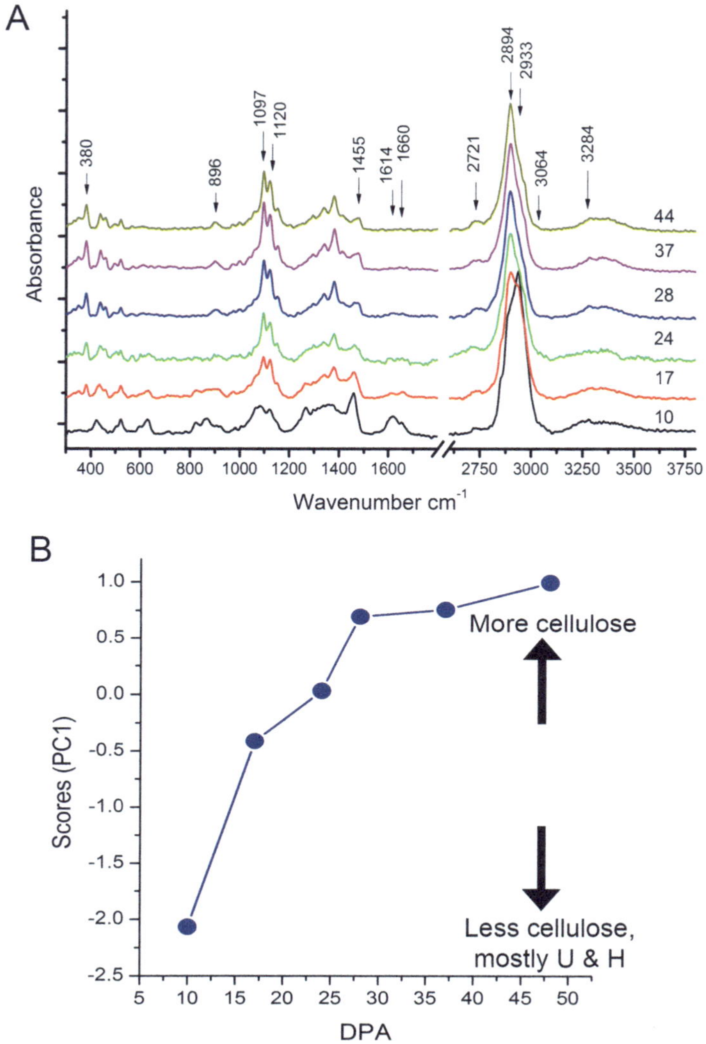

Fig. 7.4 Developmental changes in chemical compositions of *G. hirsutum* TM-1 developing fibers. (**a**) Raman spectra of TM-1 developing fibers at various DPA. (**b**) Plot of Raman principal component 1 (PC1) score versus DPA. The letters U and H indicate unsaturated hydrocarbon and long-chain alkyl groups, respectively. The figures were modified from Kim et al. (2017)

Yang et al. 2014) as well as very-long-chain fatty acids (VLCFAs) (Qin et al. 2007), a peptide hormone (Han et al. 2014), ROS (Mei et al. 2009), and calcium sensor (Tang et al. 2014) were reported to involved in fiber elongation process. Suppression of a putative sterol carrier reducing plasmodesmal permeability activates sucrose transporter gene during fiber elongation (Zhang et al. 2017). Extensive lists of cotton genes regulating fiber elongation stage have been recently reviewed (Kim 2015; Stiff et al. 2016).

7.4 Secondary Cell Wall Cellulose Biosynthesis in *G. hirsutum* Fibers

Crystalline cellulose is synthesized during the SCW stage and eventually composed of most mass of mature fibers. The cellulose properties such as amount, content, crystallinity, orientation, degree of packing, and molecular weight (MW) distribution affect fiber properties including maturity responsible for dye uptake, fineness determining the number of cotton fibers in a yarn, and breaking force affecting spinning process. Despite the importance, the SCW biosynthesis of cotton fibers was relatively less characterized than fiber elongation process due to difficulties of measuring the degree of fiber cellulose deposition and extracting high quality of RNAs from developing fibers older than 28 DPA with high cellulose contents (>75%).

The transcript levels involved in the SCW cellulose biosynthesis dramatically upregulate at the transition stage (14–16 DPA), and the increased level of cellulose content of the developing fibers at 17 DPA is detected by a chemical analysis (Updegraff 1969) (Fig. 7.5). The cellulose content gradually increases to 90–95% of the developing fibers at 37 DPA.

The chemical composition analyses using FT-IR or Raman spectroscopy can distinguish the SCW cellulose from the PCW cellulose (Kim et al. 2017). The positive PC1 scores in the Raman loading plot (Fig. 7.4b) show that the developing fibers from 24 to 44 DPA consist of SCW cellulose which is the highly ordered structures of the β-1,4-glucan chains and called cellulose microfibrils (CMFs). The developing fibers at the transition stage (17 DPA) show a SCW cellulose peak at 380 cm^{-1} (Agarwal et al. 2010) that attributed to the crystalline cellulose (Fig. 7.4a).

The molecular weight of SCW cellulose is greater than that in the PCW cellulose (Marx-Figini 1982; Timpa and Triplett 1993), and the SCW cellulose contains more numbers of glucose residues per cellulose molecule than the PCW cellulose (Delmer

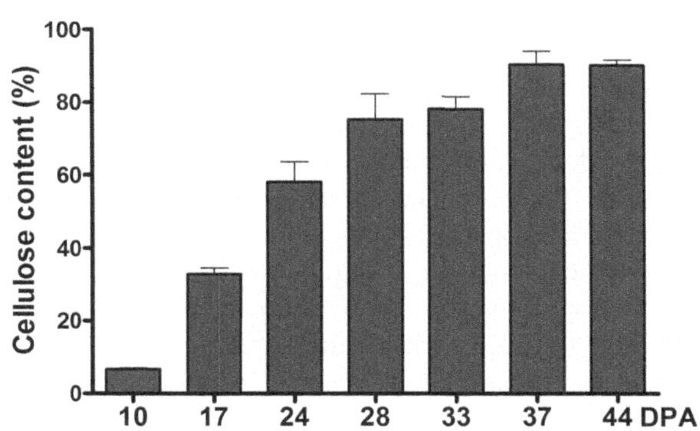

Fig. 7.5 Fiber cellulose biosynthesis during fiber development. Average cellulose contents of developing fibers at various DPAs from *G. hirsutum* TM-1 were measured. The figures were modified from Kim et al. (2017)

and Amor 1995; Timpa and Triplett 1993). Cellulose is synthesized by a large membrane-bound protein complex that is named as cellulose synthesis complex (CSC) comprised of multiple cellulose synthase proteins (CESAs) and other integral proteins (Doblin et al. 2002). The first plant *CESAs* (a full length *GhCesA1* and a 5'-truncated partial *GhCesA2*) were identified from developing cotton fibers during the SCW stage (Pear et al. 1996). The 3D computational model of a cotton GhCESA1 protein (Sethaphong et al. 2013) and *Arabidopsis* CESAs (Sethaphong et al. 2016; Slabaugh et al. 2014) was predicted based on the crystal structure of bacterial cellulose synthase (BcsA) from *Rhodobacter sphaeroides* (Morgan et al. 2013). According to the recent "hexamer of trimers" model, a CSC is composed of 18 CESA proteins with 3 CESAs per lobe as the mostly likely composition of a rosette to account for an 18-chain CMF (Gonneau et al. 2014; Hill et al. 2014; Newman et al. 2013; Vandavasi et al. 2016). Upland cotton TM-1 genome contains at least 32 *GhCESA* family genes (Zhang et al. 2015). They are classified into two major groups: one is expressed during PCW development, and the other including at least 10 *GhCESA*s is predominately expressed during SCW biosynthesis. In addition, genes encoding a *sucrose synthase* (Brill et al. 2011), a *chitinase-like* (Zhang et al. 2004), *WLIM1a* (Han et al. 2013), and *GhADF1*(Wang et al. 2009) were reported to involve in SCW biosynthesis process of cotton fibers.

As cotton fibers are thickening during the SCW stage, average cell wall area measured from the cross-sectioned fiber proportionally increased and reached to approximately 150 μm^2 at 44 DPA, whereas average perimeter (approximately 55 μm) of developing fibers is not changed (Fig. 7.6a). The HVI micronair (MIC) values, which indirectly measure a combination of fiber maturity and fineness, proportionally increase during the SCW thickening stage (Fig. 7.6b). The MIC is an indirect measurement of the air permeability through a mass of fiber compressed to a fixed volume. The degree of fiber wall thickening is directly measured from the cross-sectioned fibers by calculating circularity ($\theta = 4\pi A/P^2$; A, wall area; P, perimeter) and then converting the θ values to fiber maturity ratio (MR) values (Thibodeaux and Evans 1986). The calculated MR values of the developing fibers during the SCW stage were correlated with the corresponding MIC values (Fig. 7.6b).

Crystallinity of the SCW cellulose has been studied with polarized microscopes and X-ray diffraction (XRD) in qualitative or semiquantitative ways (Kim et al. 2017). The polarized microscopic images (Fig. 7.7) show that birefringence that is a measure of CMF organization increases in the fibers as the amount of crystalline cellulose increases in the fiber SCW. The developing fibers with less crystalline cellulose at 24 DPA appear blue color due to lower intensity of birefringence, whereas those with more crystalline cellulose at 44 DPA appear yellow color due to higher intensity of birefringence generated from organized crystalline cellulose (Fig. 7.7). The CMFs are arranged helically around the growing fiber with periodic reversals in the deposition angle (Seagull 1986, 1993). Between 28 and 33 DPA, developing fibers become twisted due to the reversal regions where CMF orientation changes (Haigler 2010; Kim and Triplett 2001).

The orientation of CMFs in developing fiber correlates with fiber strength (Moharir 1998; Moharir et al. 1999). Attenuated total reflection Fourier transform

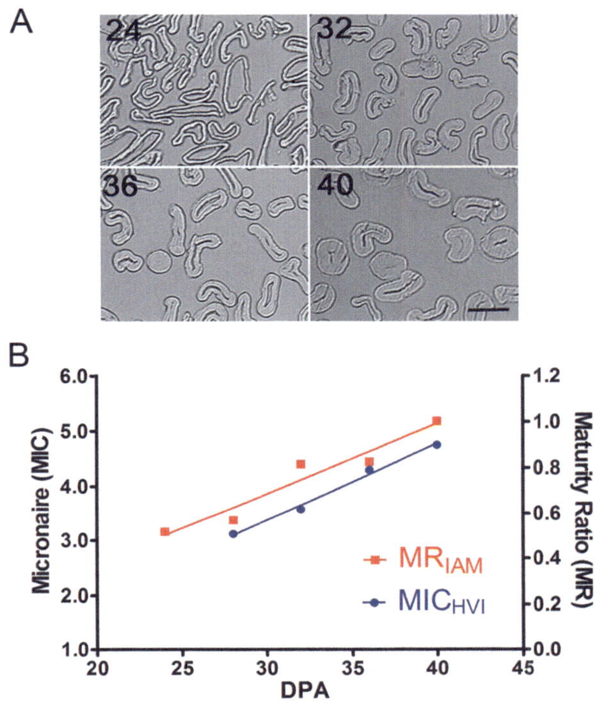

Fig. 7.6 Measurements of fiber maturity representing the degree of fiber wall thickness during secondary cell wall biosynthesis stage from *G. hirsutum* TM-1 fibers. (**a**) Fiber cross-sectional image analysis microscopy (IAM) of developing fibers at different developmental stages (24–40 DPA). A scale bar represents 10 μm. (**b**) Monitoring of maturity ratio (MR) values from developing fibers by IAM method and Micronare (MIC) values by high-volume instrument (HVI). The figures were modified from Kim et al. (2014)

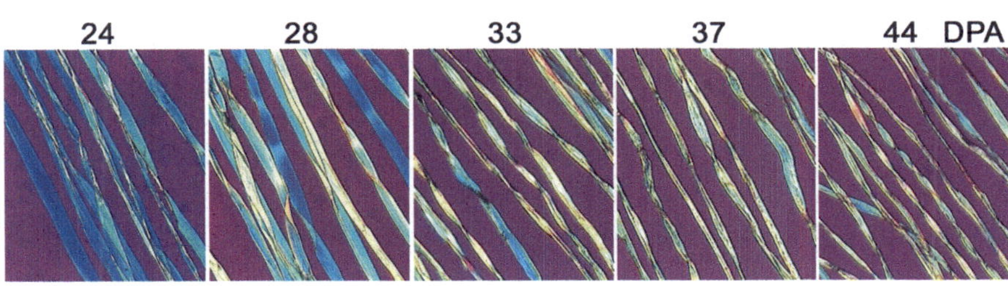

Fig. 7.7 Comparisons of polarized microscopic images of *G. hirsutum* TM-1 fibers at secondary cell wall stage. Developing fibers (24–37 DPA) and fully developed fibers (44 DPA) were photographed by polarized microscopy. Less mature fibers appeared blue color, whereas more mature fibers appeared yellow color that represents birefringence generated from organized crystalline cellulose. The figures were modified from Kim et al. (2017)

infrared (ATR FT-IR) spectroscope measurement (Kim et al. 2018; Liu et al. 2012) shows that the bundle strength values of developing fibers increase as the crystallinity (CI_{IR}) values increase during the SCW biosynthesis stage (Fig. 7.8). Comparative transcriptome analyses of two NILs differing in single fiber breaking forces suggested that receptor-like kinase (RLK) signaling pathways might be involved in cell wall integrity and strength (Islam et al. 2016).

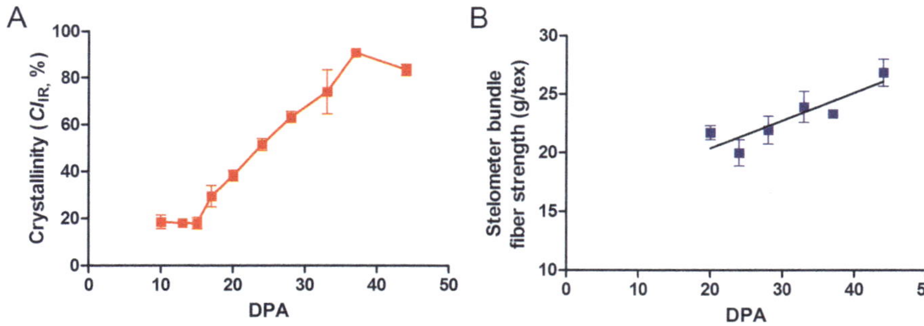

Fig. 7.8 Measurements of crystallinity and fiber strength during fiber development of *G. hirsutum* MD52ne. (**a**) Average crystallinity (CI_{IR}) values were determined from six replicates of developing fibers at ten different developmental time points (10, 13, 15, 17, 20, 24, 28, 33, 37, and 44 DPA) by ATR FT-IR spectroscopy. (**b**) Average values of bundle fiber strength from developing fibers (20, 24, 28, 33, 37, and 44 DPA) were measured from three replicates by Stelometer. The figures were modified from Islam et al. (2016)

The lateral order of CMFs was associated with the degree of fiber SCW thickness (Kim et al. 2017, 2013a). The immature fiber (*im*) mutant with low fiber maturity and its isogenic wild-type line TM-1 with high fiber maturity (Fig. 7.9a, b) have become a model system for studying cotton fiber thickness (Kim et al. 2013a, b). A pentatricopeptide repeat (PPR) gene was linked to the *im* fiber phenotype (Thyssen et al. 2016). Sum-frequency generation (SFG) vibration spectroscopy measurement showed the significant CMF packing differences between the TM-1 and *im* mutant differing in fiber maturity (Fig. 7.9c). The mesoscale spatial packing of cellulose crystallites influences the SFG intensity ratio of the peaks corresponding to alkyl (CH and CH_2) and hydroxyl (OH) stretching vibrations (Lee et al. 2015, 2014). The lower CH_2/OH intensity ratio of developing *im* fibers over developing TM-1 fibers implies the disruption of CMF organization or assembly in the SCW of the *im* fibers as compared to the TM-1 fibers (Kim et al. 2017). The results support a notion that the CMF packing is a potential factor determining fiber maturity.

7.5 Maturation Process in *G. hirsutum* Fiber

Most upland cotton varieties grown in regular US cotton fields reach their full maturity at 40–50 DPA (Kim 2015; Snider and Oosterhuis 2015). Several reports showed that fiber development of upland cotton is delayed when they are grown under environmental stress (Bradow and Davidonis 2000) or pots (Abidi et al. 2010, 2008; Lee et al. 2015). During the maturation stage from the developing TM-1 fibers at 37 DPA to the fully developed TM-1 fibers at 44 DPA, the alkyl/hydroxyl area ratios of the SFG vibrational intensities continue increasing (Fig. 7.9c) although there are little changes of the cellulose content (90–95%) (Fig. 7.5).These results strongly

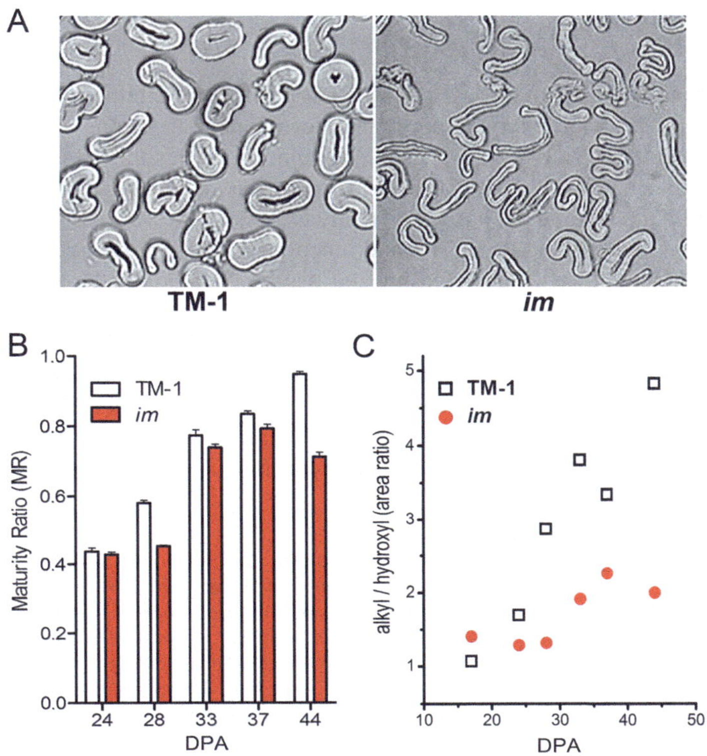

Fig. 7.9 Relationships between fiber thickness and cellulose microfibril assembly. (**a**) Comparisons of cross sections from fully developed fibers (44 DPA) between *G. hirsutum* wild-type TM-1 and its near isogenic *im* mutant defective in fiber cell wall thickness. (**b**) Comparisons of fiber maturity ratio (MR) values measured by Cottonscope from developing fibers (24, 28, 33, and 37 DPA) and developed fibers (44 DPA) of the NILs. (**c**) Comparisons of the CMF assembly measured by sum-frequency generation (SFG) vibration spectroscopy from the NIL fibers. Alkyl/hydroxyl ratio SFG area ratio for TM-1 (black square) and *im* (red circle) indicates the lateral CMF orders of TM-1 and *im* during cotton fiber development. The figures were modified from Kim et al. (2017)

imply that the CMF organization or assembly of upland TM-1 fibers is organizing during the maturation stage. As the fibers dehisce at maturity, the cytoplasm adheres to the innermost layer of the fiber cell wall, the vacuole collapses, and the fiber changes from a cylindrical shape to a bean shape (Fig. 7.9a). Immature fibers among the developed fibers cause problems for the textile industry such as yarn breakage and uneven dye uptake. Fully developed fibers mainly consist of cellulose Iβ that is mostly found in higher plants (Atalla and Vanderhart 1984). In contrast to the morphological changes of cotton fibers at the maturation stage that can be easily observed, the molecular mechanisms regulating the fiber maturation process have been little understood due to difficulties of extracting high-quality RNA from the mature fiber cells.

7.6 Conclusion

Cotton fibers are mainly composed of secondary cell wall cellulose. Cellulose biosynthesis and deposition during cotton fiber development determine fiber maturity, fineness, and strength as well as yield. With innovative genomics techniques and cotton genome sequences, the list of candidate genes regulating cellulose biosynthesis and fiber development has been greatly expanded. However, there have been challenges of determining their roles and functions with the conventional HVI and AFIS instruments that do not measure the degree of cellulose biosynthesis and deposition from cotton fibers. In this review chapter, the author described upland cotton fiber development using both conventional and advanced phenotypic tools. Reliable and accurate phenotypic data of developing and developed fibers can be used for bridging the gap between genotypic and phenotypic cotton researches, interpreting transcriptome profile data sets that are rapidly advancing with emerging techniques, and providing a way of dissecting functions of candidate genes involved in fiber development.

Acknowledgments This research was supported by the USDA-ARS CRIS Project # 6054-21000-017-00D and Cotton Incorporated-sponsored project #12-199. Author thanks Dr. Rick Turley of USDA-ARS, SEA, for providing constructive feedback on the manuscript. Author acknowledges contributions from former and current members of the Cotton Fiber Bioscience Research Unit and the Cotton Structure & Quality Research Unit of Southern Regional Research Center, USDA-ARS, and collaborators from Pennsylvania State University for Raman and SFG vibrational spectroscopic analyses. Mention of trade names or commercial products in this publication is solely for the purpose of providing specific information and does not imply recommendation or endorsement by the USDA that is an equal opportunity employer.

References

Abidi N, Cabrales L, Hequet E (2010) Fourier transform infrared spectroscopic approach to the study of the secondary cell wall development in cotton fiber. Cellulose 17:309–320

Abidi N, Hequet E, Cabrales L, Gannaway J, Wilkins T, Wells LW (2008) Evaluating cell wall structure and composition of developing cotton fibers using Fourier transform infrared spectroscopy and thermogravimetric analysis. J Appl Polym Sci 107:476–486

Agarwal UP, Reiner RS, Ralph SA (2010) Cellulose I crystallinity determination using FT–Raman spectroscopy: univariate and multivariate methods. Cellulose 17:721–733

Aleman L, Kitamura J, Abdel-Mageed H, Lee J, Sun Y, Nakajima M, Ueguchi-Tanaka M, Matsuoka M, Allen RD (2008) Functional analysis of cotton orthologs of GA signal transduction factors GID1 and SLR1. Plant Mol Biol 68:1–16

Alleman L, Allen RD (2010) Research in cotton fibre improvement. In: Singh B (ed) Industrial crops and uses. CAB International, Cambridge, pp 277–307

An C, Saha S, Jenkins JN, Scheffler BE, Wilkins TA, Stelly DM (2007) Transcriptome profiling, sequence characterization, and SNP-based chromosomal assignment of the EXPANSIN genes in cotton. Mol Gen Genomics 278:539–553

Applequist WL, Cronn R, Wendel JF (2001) Comparative development of fiber in wild and cultivated cotton. Evol Dev 3:3–17

Atalla RH, Vanderhart DL (1984) Native cellulose: a composite of two distinct crystalline forms. Science 223:283–285

Avci U, Pattathil S, Singh B, Brown VL, Hahn MG, Haigler CH (2013) Cotton fiber cell walls of Gossypium hirsutum and Gossypium barbadense have differences related to loosely-bound xyloglucan. PLoS One 8:e56315

Basra AS, Malik C (1984) Development of the cotton fiber. Int Rev Cytol 89:65–113

Beasley C, Ting IP (1974) Effects of plant growth substances on in vitro fiber development from unfertilized cotton ovules. Am J Bot 61:188–194

Benedict CR, Kohel JR, Lewis HL (1999) Cotton fiber quality. In: Smith CW, Cothren JT (eds) Cotton: origin, history, technology, and production. Wiley, New York, pp 269–288

Bradow JM, Davidonis GH (2000) Quantitation of fiber quality and the cotton production-processing interface: a physiologist's perspective. J Cotton Sci 4:34–64

Brill E, van Thournout M, White RG, Llewellyn D, Campbell PM, Engelen S, Ruan YL, Arioli T, Furbank RT (2011) A novel isoform of sucrose synthase is targeted to the cell wall during secondary cell wall synthesis in cotton fiber. Plant Physiol 157:40–54

Delmer DP, Amor Y (1995) Cellulose biosynthesis. Plant Cell 7:987–1000

Deng F, Tu L, Tan J, Li Y, Nie Y, Zhang X (2011) GbPDF1 is involved in cotton fiber initiation via the core cis-element HDZIP2ATATHB2. Plant Physiol 158:890–904

Doblin MS, Kurek I, Jacob-Wilk D, Delmer DP (2002) Cellulose biosynthesis in plants: from genes to rosettes. Plant Cell Physiol 43:1407–1420

Gonneau M, Desprez T, Guillot A, Vernhettes S, Hofte H (2014) Catalytic subunit stoichiometry within the cellulose synthase complex. Plant Physiol 166:1709–1712

Guan X, Chen ZJ (2013) Cotton fiber genomics. In: Becraft P (ed) Seed genomics. Wiley, Hoboken, pp 203–216

Haigler C (2010) Physiological and anatomical factors determining fiber structure and utility. In: Stewart JM, Oosterhuis DM, Heitholt JJ, Mauney JR (eds) Physiology of cotton. Springer, New York, pp 33–47

Haigler CH, Betancur L, Stiff MR, Tuttle JR (2012) Cotton fiber: a powerful single-cell model for cell wall and cellulose research. Front Plant Sci 3:1–7

Han J, Tan J, Tu L, Zhang X (2014) A peptide hormone gene, GhPSK promotes fibre elongation and contributes to longer and finer cotton fibre. Plant Biotechnol J 12:861–871

Han LB, Li YB, Wang HY, Wu XM, Li CL, Luo M, Wu SJ, Kong ZS, Pei Y, Jiao GL, Xia GX (2013) The dual functions of WLIM1a in cell elongation and secondary wall formation in developing cotton fibers. Plant Cell 25:4421–4438

Hill JL, Hammudi MB, Tien M (2014) The Arabidopsis cellulose synthase complex: a proposed hexamer of CESA trimers in an equimolar stoichiometry. Plant Cell 26:4834–4842

Hu H, He X, Tu L, Zhu L, Zhu S, Ge Z, Zhang X (2016) GhJAZ2 negatively regulates cotton fiber initiation by interacting with the R2R3-MYB transcription factor GhMYB25-like. Plant J 88:921–935

Islam MS, Fang DD, Thyssen GN, Delhom CD, Liu Y, Kim HJ (2016) Comparative fiber property and transcriptome analyses reveal key genes potentially related to high fiber strength in cotton (Gossypium hirsutum L.) line MD52ne. BMC Plant Biol 16:36

Kearney TH, Harrison GJ (1927) Inheritance of smooth seeds in cotton. J Agric Res 35:193–217

Kim HJ (2015) Fiber biology. In: Fang DD, Percy RG (eds) Cotton, 2nd edn. American Society of Agronomy, Crop Science Society of America, and Soil Science Society of America, Madison, pp 97–127

Kim HJ, Hinchliffe D, Triplett B, Yeater K, Moon H, Gilbert M, Thyssen G, Turley R, Fang D (2015) Phytohormonal networks promote differentiation of fiber initials on pre-anthesis cotton ovules grown in vitro and in planta. PLoS One 10(4):e0125046

Kim HJ, Lee CM, Dazen K, Delhom CD, Liu Y, Rodgers JE, French AD, Kim SH (2017) Comparative physical and chemical analyses of cotton fibers from two near isogenic upland lines differing in fiber wall thickness. Cellulose 24:2385–2401

Kim HJ, Liu Y, French AD, Lee CM, Kim SH (2018) Comparison and validation of Fourier transform infrared spectroscopic methods for monitoring secondary cell wall cellulose from cotton fibers. Cellulose 25:49–64
Kim HJ, Moon HS, Delhom CD, Zeng L, Fang DD (2013a) Molecular markers associated with the immature fiber (im) gene affecting the degree of fiber cell wall thickening in cotton (Gossypium hirsutum L.). Theor Appl Genet 126:23–31
Kim HJ, Rodgers J, Delhom C, Cui X (2014) Comparisons of methods measuring fiber maturity and fineness of upland cotton fibers containing different degree of fiber cell wall development. Text Res J 84:1622–1633
Kim HJ, Tang Y, Moon HS, Delhom CD, Fang DD (2013b) Functional analyses of cotton (Gossypium hirsutum L.) immature fiber (im) mutant infer that fiber cell wall development is associated with stress responses. BMC Genomics 14:889
Kim HJ, Triplett BA (2001) Cotton fiber growth in planta and in vitro. Models for plant cell elongation and cell wall biogenesis. Plant Physiol 127:1361–1366
Kohel R, Richmond T, Lewis C (1970) Texas marker-1. Description of a genetic standard for Gossypium hirsutum L. Crop Sci 10:670–671
Lang AG (1938) The origin of lint and fuzz hairs of cotton. J Agric Res 56:507–521
Lee CM, Kafle K, Belias DW, Park YB, Glick RE, Haigler CH, Kim SH (2015) Comprehensive analysis of cellulose content, crystallinity, and lateral packing in Gossypium hirsutum and Gossypium barbadense cotton fibers using sum frequency generation, infrared and Raman spectroscopy, and X-ray diffraction. Cellulose 22:971–989
Lee CM, Kafle K, Park YB, Kim SH (2014) Probing crystal structure and mesoscale assembly of cellulose microfibrils in plant cell walls, tunicate tests, and bacterial films using vibrational sum frequency generation (SFG) spectroscopy. Phys Chem Chem Phys 16:10844–10853
Lee JJ, Hassan OS, Gao W, Wei NE, Kohel RJ, Chen X-Y, Payton P, Sze S-H, Stelly DM, Chen ZJ (2006) Developmental and gene expression analyses of a cotton naked seed mutant. Planta 223:418–432
Lee JJ, Woodward AW, Chen ZJ (2007) Gene expression changes and early events in cotton fibre development. Ann Bot 100:1391–1401
Li C, Guo W, Zhang T (2009) Fiber initiation development in upland cotton (Gossypium hirsutum L.) cultivars varying in lint percentage. Euphytica 165:223–230
Li F, Fan G, Lu C, Xiao G, Zou C, Kohel RJ, Ma Z, Shang H, Ma X, Wu J (2015) Genome sequence of cultivated upland cotton (Gossypium hirsutum TM-1) provides insights into genome evolution. Nat Biotechnol 33:524–530
Li F, Fan G, Wang K, Sun F, Yuan Y, Song G, Li Q, Ma Z, Lu C, Zou C (2014) Genome sequence of the cultivated cotton Gossypium arboreum. Nat Genet 46:567–572
Li X-B, Fan X-P, Wang X-L, Cai L, Yang W-C (2005) The cotton ACTIN1 gene is functionally expressed in fibers and participates in fiber elongation. Plant Cell 17:859–875
Liao W-b, Ruan M-b, Cui B-m, Xu N-f, Lu J-j, Peng M (2009) Isolation and characterization of a GAI/RGA-like gene from Gossypium hirsutum. Plant Growth Regul 58:35–45
Liu X, Zhao B, Zheng H-J, Hu Y, Lu G, Yang C-Q, Chen J-D, Chen J-J, Chen D-Y, Zhang L (2015) Gossypium barbadense genome sequence provides insight into the evolution of extra-long staple fiber and specialized metabolites. Sci Rep 5:14139
Liu Y, Kim HJ (2017) Fourier transform infrared spectroscopy (FT-IR) and simple algorithm analysis for rapid and non-destructive assessment of developmental cotton fibers. Sensors 17:1469
Liu Y, Thibodeaux D, Gamble G, Bauer P, VanDerveer D (2012) Comparative investigation of Fourier transform infrared (FT-IR) spectroscopy and X-ray diffraction (XRD) in the determination of cotton fiber crystallinity. Appl Spectrosc 66:983–986
Luo M, Xiao Y, Li X, Lu X, Deng W, Li D, Hou L, Hu M, Li Y, Pei Y (2007) GhDET2, a steroid 5α-reductase, plays an important role in cotton fiber cell initiation and elongation. Plant J 51:419–430
Ma Q-F, Wu C-H, Wu M, Pei W-F, Li X-L, Wang W-K, Zhang J, Yu J-W, Yu S-X (2016) Integrative transcriptome, proteome, phosphoproteome and genetic mapping reveals new aspects in a fiberless mutant of cotton. Sci Rep 6:24485

Machado A, Wu Y, Yang Y, Llewellyn DJ, Dennis ES (2009) The MYB transcription factor GhMYB25 regulates early fibre and trichome development. Plant J 59:52–62

Marx-Figini M (1982) The control of molecular weight and molecular-weight distribution in the biogenesis of cellulose. In: Cellulose and other natural polymer systems. Springer, Boston, pp 243–271

Mei W, Qin Y, Song W, Li J, Zhu Y (2009) Cotton GhPOX1 encoding plant class III peroxidase may be responsible for the high level of reactive oxygen species production that is related to cotton fiber elongation. J Genet Genomics 36:141–150

Meinert MC, Delmer DP (1977) Changes in biochemical composition of the cell wall of the cotton fiber during development. Plant Physiol 59:1088–1097

Moharir AV (1998) True-spiral angle in diploid and tetraploid native cotton fibers grown at different locations. J Appl Polym Sci 70:303–310

Moharir AV, Van Langenhove L, Van Nimmen E, Louwarie J, Kiekens P (1999) Stability of x-ray cellulose crystallite orientation parameters in native cotton with change of location and year of growth. J Appl Polym Sci 72:269–276

Morgan JL, Strumillo J, Zimmer J (2013) Crystallographic snapshot of cellulose synthesis and membrane translocation. Nature 493:181–186

Naithani S, Rao NR, Singh Y (1982) Physiological and biochemical changes associated with cotton fibre development. Physiol Plant 54:225–229

Newman RH, Hill SJ, Harris PJ (2013) Wide-angle x-ray scattering and solid-state nuclear magnetic resonance data combined to test models for cellulose microfibrils in mung bean cell walls. Plant Physiol 163:1558–1567

Pang M, Woodward AW, Agarwal V, Guan X, Ha M, Ramachandran V, Chen X, Triplett BA, Stelly DM, Chen ZJ (2009) Genome-wide analysis reveals rapid and dynamic changes in miRNA and siRNA sequence and expression during ovule and fiber development in allotetraploid cotton (Gossypium hirsutum L.). Genome Biol 10:R122

Paterson AH, Wendel JF, Gundlach H, Guo H, Jenkins J, Jin D, Llewellyn D, Showmaker KC, Shu S, Udall J, Yoo M-J, Byers R, Chen W, Doron-Faigenboim A, Duke MV, Gong L, Grimwood J, Grover C, Grupp K, Hu G, Lee T-H, Li J, Lin L, Liu T, Marier BS, Page JT, Roberts AW, Romanel E, Sanders WS, Szadkowski E, Tan X, Tang H, Xu C, Wang J, Wang Z, Zhang D, Zhang L, Ashrafi H, Bedon F, Bowers JE, Brubaker CL, Chee PW, Das S, Gingle AR, Haigler CH, Harker D, Hoffmann LV, Hovav R, Jones DC, Lemke C, Mansoor S, Rahman MU, Rainville LN, Rambani A, Reddy UK, Rong J-K, Saranga Y, Scheffler BE, Scheffler JA, Stelly DM, Triplett BA, Van Deynze A, Vaslin MFS, Waghmare VN, Walford SA, Wright RJ, Zaki EA, Zhang T, Dennis ES, Mayer KFX, Peterson DG, Rokhsar DS, Wang X, Schmutz J (2012) Repeated polyploidization of Gossypium genomes and the evolution of spinnable cotton fibres. Nature 492:423–427

Paudel D, Hequet E, Noureddine A (2013) Evaluation of cotton fiber maturity measurements. Ind Crop Prod 45:435–441

Pear JR, Kawagoe Y, Schreckengost WE, Delmer DP, Stalker DM (1996) Higher plants contain homologs of the bacterial celA genes encoding the catalytic subunit of cellulose synthase. PNAS 93:12637–12642

Pu L, Li Q, Fan X, Yang W, Xue Y (2008) The R2R3 MYB transcription factor GhMYB109 is required for cotton fiber development. Genetics 180:811–820

Qin Y-M, Hu C-Y, Pang Y, Kastaniotis AJ, Hiltunen JK, Zhu Y-X (2007) Saturated very-long-chain fatty acids promote cotton fiber and Arabidopsis cell elongation by activating ethylene biosynthesis. Plant Cell 19:3692–3704

Qin Y-M, Zhu Y-X (2011) How cotton fibers elongate: a tale of linear cell-growth mode. Curr Opin Plant Biol 14:106–111

Ruan Y-L, Llewellyn DJ, Furbank RT (2001) The control of single-celled cotton fiber elongation by developmentally reversible gating of plasmodesmata and coordinated expression of sucrose and K+ transporters and expansin. Plant Cell 13:47–60

Ruan Y-L, Llewellyn DJ, Furbank RT (2003) Suppression of sucrose synthase gene expression represses cotton fiber cell initiation, elongation, and seed development. Plant Cell 15:952–964

Ruan Y-L, Xu S-M, White R, Furbank RT (2004) Genotypic and developmental evidence for the role of plasmodesmatal regulation in cotton fiber elongation mediated by callose turnover. Plant Physiol 136:4104–4113

Ryser U (1999) Cotton fiber initiation and histodifferentiation. In: Cotton fibers: developmental biology, quality improvement, and textile processing. The Haworth Press, New York, pp 1–46

Seagull RW (1986) Changes in microtubule organization and wall microfibril orientation during in vitro cotton fiber development: an immunofluorescent study. Can J Bot 64:1373–1381

Seagull RW (1993) Cytoskeletal involvement in cotton fiber growth and development. Micron 24:643–660

Sethaphong L, Davis JK, Slabaugh E, Singh A, Haigler CH, Yingling YG (2016) Prediction of the structures of the plant-specific regions of vascular plant cellulose synthases and correlated functional analysis. Cellulose 23:145–161

Sethaphong L, Haigler CH, Kubicki JD, Zimmer J, Bonetta D, DeBolt S, Yingling YG (2013) Tertiary model of a plant cellulose synthase. PNAS 110:7512–7517

Shi Y-H, Zhu S-W, Mao X-Z, Feng J-X, Qin Y-M, Zhang L, Cheng J, Wei L-P, Wang Z-Y, Zhu Y-X (2006) Transcriptome profiling, molecular biological, and physiological studies reveal a major role for ethylene in cotton fiber cell elongation. Plant Cell 18:651–664

Singh B, Cheek HD, Haigler CH (2009) A synthetic auxin (NAA) suppresses secondary wall cellulose synthesis and enhances elongation in cultured cotton fiber. Plant Cell Rep 28:1023–1032

Slabaugh E, Sethaphong L, Xiao C, Amick J, Anderson CT, Haigler CH, Yingling YG (2014) Computational and genetic evidence that different structural conformations of a non-catalytic region affect the function of plant cellulose synthase. J Exp Bot 65:6645–6653

Snider JL, Oosterhuis DM (2015) Physiology. In: Fang DD, Percy RG (eds) Cotton, 2nd edn. American Society of Agronomy, Crop Science Society of America, and Soil Science Society of America, Madison, pp 339–400

Stewart JM (1975) Fiber initiation on the cotton ovule (Gossypium hirsutum). Am J Bot 62:723–730

Stiff MR, Haigler CH (2012) Recent advances in cotton fiber development. In: Oosterhuis DM, Cothren JT (eds) Flowering and fruiting in cotton. The Cotton Foundation, Cordova, pp 163–192

Stiff MR, Tuttle JR, Graham BP, Haigler CH (2016) Cotton fiber biotechnology cotton fiber biotechnology: potential controls and transgenic improvement of elongation and cell wall thickening. In: Ramawat KG, Ahuja MR (eds) Fiber plants biology, biotechnology and applications. Springer, Switzerland, pp 127–153

Tang W, Tu L, Yang X, Tan J, Deng F, Hao J, Guo K, Lindsey K, Zhang X (2014) The calcium sensor GhCaM7 promotes cotton fiber elongation by modulating reactive oxygen species (ROS) production. New Phytol 202:509–520

Thibodeaux DP, Evans JP (1986) Cotton fiber maturity by image analysis. Text Res J 56:130–139

Thyssen GN, Fang DD, Turley RB, Florane CB, Li P, Mattison CP, Naoumkina M (2017) A Gly65Val substitution in an actin, GhACT_LI1, disrupts cell polarity and F-actin organization resulting in dwarf, lintless cotton plants. Plant J 90:111–121

Thyssen GN, Fang DD, Zeng L, Song X, Delhom CD, Condon TL, Li P, Kim HJ (2016) The immature fiber mutant phenotype of cotton (Gossypium hirsutum) is linked to a 22-bp frameshift deletion in a mitochondria targeted pentatricopeptide repeat gene. G3: Genes, Genomes, Genetics 6:1627–1633

Timpa JD, Triplett BA (1993) Analysis of cell-wall polymers during cotton fiber development. Planta 189:101–108

Turley R, Kloth R (2002) Identification of a third fuzzless seed locus in upland cotton (Gossypium hirsutum L.). J Hered 93:359–364

Turley RB, Kloth RH (2008) The inheritance model for the fiberless trait in upland cotton (Gossypium hirsutum L.) line SL1-7-1: variation on a theme. Euphytica 164:123–132

Udall JA (2009) The gossypium transcriptome. In: Genetics and genomics of cotton. Springer, New York, pp 157–185

Updegraff DM (1969) Semimicro determination of cellulose in biological materials. Anal Biochem 32:420–424

Vandavasi VG, Putnam DK, Zhang Q, Petridis L, Heller WT, Nixon BT, Haigler CH, Kalluri U, Coates L, Langan P (2016) A structural study of CESA1 catalytic domain of Arabidopsis cellulose synthesis complex: evidence for CESA trimers. Plant Physiol 170:123–135

Wakelyn PJ, Bertoniere NR, French AD, Thibodeaux DP, Triplett BA, Rousselle M-A, Goynes WR Jr, Edwards JV, Hunter L, McAlister DD (2010) Cotton fiber chemistry and technology. CRC Press, Boca Raton

Walford S-A, Wu Y, Llewellyn DJ, Dennis ES (2011) GhMYB25-like: a key factor in early cotton fibre development. Plant J 65:785–797

Walford S-A, Wu Y, Llewellyn DJ, Dennis ES (2012) Epidermal cell differentiation in cotton mediated by the homeodomain leucine zipper gene, GhHD-1. Plant J 71:464–478

Wan Q, Guan X, Yang N, Wu H, Pan M, Liu B, Fang L, Yang S, Hu Y, Ye W, Zhang H, Ma P, Chen J, Wang Q, Mei G, Cai C, Yang D, Wang J, Guo W, Zhang W, Chen X, Zhang T (2016) Small interfering RNAs from bidirectional transcripts of GhMML3_A12 regulate cotton fiber development. New Phytol 210:1298–1310

Wang HY, Wang J, Gao P, Jiao GL, Zhao PM, Li Y, Wang GL, Xia GX (2009) Down-regulation of GhADF1 gene expression affects cotton fibre properties. Plant Biotechnol J 7:13–23

Wang L, Cook A, Patrick JW, Chen X-Y, Ruan Y-L (2014) Silencing the vacuolar invertase gene GhVIN1 blocks cotton fiber initiation from the ovule epidermis, probably by suppressing a cohort of regulatory genes via sugar signaling. Plant J 78:686–696

Ware JO, Benedict LI, Rolfe WH (1947) A recessive naked seed character in upland cotton. J Hered 38:313–319

Wendel JF, Flagel LE, Adams KL (2012) Jeans, genes, and genomes: cotton as a model for studying polyploidy. In: Polyploidy and genome evolution. Springer, New York, pp 181–207

Wu H, Tian Y, Wan Q, Fang L, Guan X, Chen J, Hu Y, Ye W, Zhang H, Guo W, Chen X, Zhang T (2017) Genetics and evolution of MIXTA genes regulating cotton lint fiber development. New Phytol 217:883–895

Wu Y, Llewellyn DJ, White R, Ruggiero K, Al-Ghazi Y, Dennis ES (2007) Laser capture microdissection and cDNA microarrays used to generate gene expression profiles of the rapidly expanding fibre initial cells on the surface of cotton ovules. Planta 226:1475–1490

Wu Y, Machado AC, White RG, Llewellyn DJ, Dennis ES (2006) Expression profiling identifies genes expressed early during lint fibre initiation in cotton. Plant Cell Physiol 47:107–127

Xiao Y-H, Li D-M, Yin M-H, Li X-B, Zhang M, Wang Y-J, Dong J, Zhao J, Luo M, Luo X-Y (2010) Gibberellin 20-oxidase promotes initiation and elongation of cotton fibers by regulating gibberellin synthesis. J Plant Physiol 167:829–837

Xie F, Wang Q, Zhang B (2015) Global microRNA modification in cotton (Gossypium hirsutum L.). Plant Biotechnol J 13:492–500

Yang SS, Cheung F, Lee JJ, Ha M, Wei NE, Sze SH, Stelly DM, Thaxton P, Triplett B, Town CD (2006) Accumulation of genome-specific transcripts, transcription factors and phytohormonal regulators during early stages of fiber cell development in allotetraploid cotton. Plant J 47:761–775

Yang Z, Zhang C, Yang X, Liu K, Wu Z, Zhang X, Zheng W, Xun Q, Liu C, Lu L, Yang Z, Qian Y, Xu Z, Li C, Li J, Li F (2014) PAG1, a cotton brassinosteroid catabolism gene, modulates fiber elongation. New Phytol 203:437–448

Yuan D, Tang Z, Wang M, Gao W, Tu L, Jin X, Chen L, He Y, Zhang L, Zhu L (2015) The genome sequence of Sea-Island cotton (Gossypium barbadense) provides insights into the allopolyploidization and development of superior spinnable fibres. Sci Rep 5:17662

Zhang D, Hrmova M, Wan CH, Wu C, Balzen J, Cai W, Wang J, Densmore LD, Fincher GB, Zhang H, Haigler CH (2004) Members of a new group of chitinase-like genes are expressed preferentially in cotton cells with secondary walls. Plant Mol Biol 54:353–372

Zhang D, Zhang T, Guo W (2010) Effect of H2O2 on fiber initiation using fiber retardation initiation mutants in cotton (Gossypium hirsutum). J Plant Physiol 167:393–399

Zhang F, Jin X, Wang L, Li S, Wu S, Cheng C, Zhang T, Guo W (2016) GhFAnnxA affects fiber elongation and secondary cell wall biosynthesis associated with Ca2+ influx, ROS homeostasis and actin filament reorganization. Plant Physiol 171:1750–1770

Zhang H-B, Li Y, Wang B, Chee PW (2008) Recent advances in cotton genomics. Int J Plant Genomics 2008:742304

Zhang M, Zheng X, Song S, Zeng Q, Hou L, Li D, Zhao J, Wei Y, Li X, Luo M, Xiao Y, Luo X, Zhang J, Xiang C, Pei Y (2011) Spatiotemporal manipulation of auxin biosynthesis in cotton ovule epidermal cells enhances fiber yield and quality. Nat Biotechnol 29:453–458

Zhang T, Hu Y, Jiang W, Fang L, Guan X, Chen J, Zhang J, Saski CA, Scheffler BE, Stelly DM, Hulse-Kemp AM, Wan Q, Liu B, Liu C, Wang S, Pan M, Wang Y, Wang D, Ye W, Chang L, Zhang W, Song Q, Kirkbride RC, Chen X, Dennis E, Llewellyn DJ, Peterson DG, Thaxton P, Jones DC, Wang Q, Xu X, Zhang H, Wu H, Zhou L, Mei G, Chen S, Tian Y, Xiang D, Li X, Ding J, Zuo Q, Tao L, Liu Y, Li J, Lin Y, Hui Y, Cao Z, Cai C, Zhu X, Jiang Z, Zhou B, Guo W, Li R, Chen ZJ (2015) Sequencing of allotetraploid cotton (Gossypium hirsutum L. acc. TM-1) provides a resource for fiber improvement. Nat Biotechnol 33:531–537

Zhang T, Pan J (1991) Genetic analysis of a fuzzless-lintless mutant in Gossypium hirsutum L. Jiangsu J Agric Sci 7:13–16

Zhang Z, Ruan Y-L, Zhou N, Wang F, Guan X, Fang L, Shang X, Guo W, Zhu S, Zhang T (2017) Suppressing a putative sterol carrier gene reduces plasmodesmal permeability and activates sucrose transporter genes during cotton fiber elongation. Plant Cell 29:2027–2046

Chapter 8
Cotton Fiber Genes and Stable Quantitative Trait Loci

David D. Fang

8.1 Introduction

Cotton fibers are unicellular trichomes that differentiate from the epidermal cells of developing cotton ovules (Kim and Triplett 2001). Cotton fiber development lasts about 60 days and is divided into four stages: initiation, elongation, secondary cell wall (SCW) biosynthesis, and maturation (Lee et al. 2007; Qin and Zhu 2011). On 1 or 2 days before anthesis (flowering), about 25–30% of the ovule epidermal cells initiate into fiber cells. The fiber yield is positively correlated to the number of initiated fiber cells per a given area of ovule surface (Zhang et al. 2011). Once initiated, fiber cells enter the elongation stage that lasts about 3 weeks in Upland cotton (*Gossypium hirsutum* L.) and a few more days longer in Pima cotton (*Gossypium barbadense* L.) (Avci et al. 2013). During the elongation stage, fiber cells grow to 25–40 mm in Upland cotton and may reach 50 mm in Pima cotton, making them among the longest cells in the plant kingdom (Kim and Triplett 2001). Fiber length is greatly determined at the elongation stage. Around 12–16 days post anthesis (DPA), fiber cell transition from elongation to SCW biosynthesis continues until 35 DPA or later (Haigler et al. 2012; Hinchliffe et al. 2011a). At this stage, large amount of cellulose is synthesized and deposited to the SCW between the plasmalemma and the primary cell wall. Fiber strength, maturity, and fineness are largely determined at SCW biosynthesis stage. Fiber development ceases when the fruit wall dehisces and the fibers air-dry.

D. D. Fang (✉)
Cotton Fiber Bioscience Research Unit, USDA-ARS, Southern Regional Research Center, New Orleans, LA, USA
e-mail: david.fang@ars.usda.gov

D. D. Fang (ed.), *Cotton Fiber: Physics, Chemistry and Biology*,
https://doi.org/10.1007/978-3-030-00871-0_8

There are two types of fibers: lint and fuzz. Lint fiber cells initiate typically 1–2 days before the anthesis, elongate for about 3 weeks, and grow to average 30 mm in length in Upland cotton. Lint fibers can be easily separated from seeds by ginning. Fuzz fibers initiate around 3–4 DPA (Stewart 1975; Wan et al. 2016; Zhang et al. 2007) and stop elongation around 10–12 DPA. Fuzz fibers are usually less than 5 mm in length and remain attached to seeds after ginning (Zhang et al. 2007). Recent studies indicated that both lint and fuzz fibers may share common developmental pathways at least in early differentiation even though under different genetic controls (Turley and Kloth 2002; Turley and Kloth 2008; Wan et al. 2016; Wu et al. 2018; Zhu et al. 2018). It is noted that it is difficult to discern fuzz fibers from short lint fibers (<6 mm) before or after ginning.

Biological processes involved in each stage of fiber development play influential roles in the eventual fiber physical properties including length, strength, fineness, maturity, etc. It has been suggested that cotton fiber development requires concerted efforts of many genes in each biological process. During the past decades, many fiber genes have been identified, and some of them were transferred into cotton or model plants via transformation for functionality validation (Li et al. 2005; Stiff et al. 2016; Walford et al. 2011; Zhang et al. 2016). A few of these genes may be a target for genetic manipulation via biotechnology to improve fiber quality.

Besides identification of fiber genes, a lot of studies have been conducted to identify quantitative trait locus or loci (QTL) controlling fiber quality traits using methods such as biparental genetic mapping and genome-wide association study (GWAS) (Ademe et al. 2017; Du et al. 2018; Fang et al. 2014; Fang et al. 2017a; Gore et al. 2014; Huang et al. 2018; Islam et al. 2016b; Ma et al. 2018; Nie et al. 2016; Zhang et al. 2013). Many fiber QTL have been identified, and some of them are being used or ready to be used in cotton breeding via a marker-assisted selection (MAS) strategy to improve fiber quality.

In the subsequent texts of this chapter, I will provide two parts of review. The first part is on fiber gene identification and functional characterization. The goal is to have a better understanding about the molecular mechanisms of fiber development and to identify potential gene or genes that can be used for varietal improvement through genetic manipulation via biotechnology. The second part is on the fiber QTL identification. The objective is to identify important stable fiber QTL for use in breeding via MAS.

8.2 Fiber Genes

Many genes have been identified to have roles in fiber development based on transcriptomic analysis, genetic mapping, and comparative orthologous gene analysis (Hinchliffe et al. 2010; Ji et al. 2003; Lee et al. 2006; Taliercio and

Boykin 2007; Wu et al. 2006; Yang et al. 2006). Recent efforts in genome-wide RNA sequencing analysis identified many more (Ding et al. 2014; Islam et al. 2016a; Naoumkina et al. 2015; Wan et al. 2014). Altogether, there may be close to 10,000 genes that are directly or indirectly relevant to the cotton fiber development at different stages. However, only a small fraction of these genes (Table 8.1) were functionally analyzed using transformation (in cotton or other model plants such as *Arabidopsis* or tobacco), transient analysis, or virus-induced gene silencing (VIGS) (Kim et al. 2012; Machado et al. 2009; Pu et al. 2008; Stiff et al. 2016; Thyssen et al. 2017; Walford et al. 2011; Zhang et al. 2016). There are a few excellent review papers focusing on early fiber development (Lee et al. 2007), elongation (Qin and Zhu 2011), second cell wall thickening (Haigler et al. 2012), possible candidate genes for biotechnology (Haigler

Table 8.1 Fiber genes whose functions were characterized

Gene name	Fiber development stages	References
GhDEL65/61	Initiation	Shangguan et al. (2016)
GhHD1		Walford et al. (2012)
GhJAZ2		Hu et al. (2016)
GhMML3 (GhMYB25-like)		Walford et al. (2011), Wan et al. (2016), and Zhu et al. (2018)
GhMML4		Wu et al. (2018)
GaMYB2		Wang et al. (2004)
GhMYB2		Guan et al. (2014)
GhPIN1, 6, 8		Zhang et al. (2017b)
GhTTG1–2		Humphries et al. (2005)
GhVIN1		Wang et al. (2010c, 2014)
GbPDF1		Deng et al. (2012)
GbMYB2	Initiation and elongation	Huang et al. (2013b)
GhAGP4		Li et al. (2010b)
GhAnn3		Li et al. (2013a)
GhCFE1A		Lv et al. (2015)
GhGA20ox1–3		Xiao et al. (2010)
GhiaaM		Zhang et al. (2011)
GhMML7(GhMYB25)		Machado et al. (2009)
GhMYB109		Pu et al. (2008) and Suo et al. (2003)
GhSUSY		Ruan and Chourey (1998) and Ruan et al. (2001)
GhTCP14		Wang et al. (2013)

(continued)

Table 8.1 (continued)

Gene name	Fiber development stages	References
GhACO1–3	Elongation	Shi et al. (2006)
GhACTIN1		Li et al. (2005)
GhACT_LI1		Thyssen et al. (2017)
GhAnn2		Tang et al. (2014a)
GhAPX1		Li et al. (2007a)
GhATP1		He et al. (2018)
Gh-BtubL		Ji et al. (2002)
GhCaM7		Tang et al. (2014b)
GhCDPK1		Zhang et al. (2015a)
GhCPC		Liu et al. (2015)
GhCPK1		Huang et al. (2008)
GhEXPA1		Harmer et al. (2002), Ruan et al. (2001), and Xu et al. (2013a)
GhFLA1		Huang et al. (2013a)
GhGID1a		Aleman et al. (2008)
GhGID1b		Aleman et al. (2008)
GhHOX3		Shan et al. (2014)
GhKCS13/CER6		Qin et al. (2007)
GhPEL		Wang et al. (2010a)
GhPIP2		Li et al. (2013b)
GhPOX1		Mei et al. (2009)
GhPRE1		Zhao et al. (2018)
GhRac1		Kim and Triplett (2004)
GhRDL1		Xu et al. (2013a)
GhSCP2D		Zhang et al. (2017d)
GhSLR1a		Aleman et al. (2008)
GhSLR1b		Aleman et al. (2008)
GhUGP1		Li et al. (2015)
GhTUA9		Li et al. (2007b)
GhWBC1		Zhu et al. (2003)
GhXTH1		Lee et al. (2010)
GbAnx6		Huang et al. (2013c)
GbEXPATR		Li et al. (2016)
GbTCP		Hao et al. (2012)
GhADF1	Elongation and SCW biosynthesis	Wang et al. (2009)
GhFAnnxA		Zhang et al. (2016)
GhFIM2		Zhang et al. (2017a)
GhPFN2		Wang et al. (2010b)
GhPRP5		Xu et al. (2013b)
GhWLIMa		Han et al. (2013)

(continued)

Table 8.1 (continued)

Gene name	Fiber development stages	References
GhcelA1–2	SCW biosynthesis	Pear et al. (1996)
GhFSN1		Zhang et al. (2018)
GhGDSL		Yadav et al. (2017)
GhKNL1		Gong et al. (2014)
GhPPR (Gh_A03G0489)		Thyssen et al. (2016)
GhTT2-3A		Yan et al. (2018)
GhPHYA1	Overall fiber mass	Abdurakhmonov et al. (2014)
GhSusA1		Jiang et al. (2012)

et al. 2005; Stiff et al. 2016), and effects of plant hormones on fiber development (Liao et al. 2010). Earlier review articles include those written by Kim and Triplett (2001) and Ruan (2005). Readers are referred to these papers for more information. In this paper, I will focus on the recent advancements and the gaps not covered in previous review articles. For simplicity, I organize the genes based on their putative primary functions in fiber development stages. It is noted that although maturation is one of the four stages of fiber development, little if any research regarding identification of genes related to this stage has been reported to the best of my knowledge.

8.2.1 Genes Related to Fiber Initiation

In early days when cotton researchers knew little about how cotton fiber cells initiate, they turned to *Arabidopsis* for guidance since both cotton fibers and *Arabidopsis* leaf trichomes are single-celled plant hairs. In the 1990s, *Arabidopsis* leaf trichome initiation and development were already well-understood, where at least 20 genes are required (Hulskamp et al. 1994). These include myeloblastosis (MYB) proteins GLABROUS1 (GL1), the WD40 protein TRANSPARENT TESTA GLABRA1 (TTG1), the basic helix-loop-helix protein GLABRA3 (GL3), a homeodomain protein GLABRA2 (GL2), and another MYB-like protein TRIPTYCHON (TRY) (Hulskamp 2004; Szymanski and Marks 1998). Much of the early research work on cotton fiber initiation was to identify respective trichome orthologous genes in cotton and confirm their functions in fiber development. This strategy did help cotton researchers to have gained some understandings about fiber initiation; however, it also misleads cotton researchers to some extent as significant differences exist between branched *Arabidopsis* leaf trichomes and non-branched cotton seed trichomes (fibers).

8.2.1.1 MYB Genes

In the protein complex regulating *Arabidopsis* trichome development, there are at least two MYB genes, *GL1* and *TRY*. *GL1* is a well-characterized MYB gene that plays essential role in trichome initiation in *Arabidopsis* (Larkin et al. 1994). Comparative transcriptome analysis in cotton identified many MYB genes and their possible roles in fiber development (Lee et al. 2006; Loguerico et al. 1999; Salih et al. 2016; Suo et al. 2003; Taliercio and Boykin 2007). Wang et al. (2004) isolated a *GL1* ortholog in a fiber-bearing diploid *G. arboreum*. It was a MYB gene named *GaMYB2*. Overexpression of *GaMYB2* rescued trichome formation of the *Arabidopsis gl1* mutant. In Upland cotton, the *GaMYB2* promoter exhibited activities in both developing fiber cells and in trichomes of other aerial organs (Shangguan et al. 2008). The Upland cotton MYB2 genes are regulated by microRNAs 828 and 858 (Guan et al. 2014). The MYB2 gene in *G. barbadense* affects both fiber initiation and elongation (Huang et al. 2013b). In a separate research, a MYB gene, namely, *GhMYB109*, encoding a R2R3 MYB transcription factor of 234 amino acids, was structurally related to *GL1* of *Arabidopsis*. Suppression of *GhMYB109* led to a substantial reduction of cytoskeleton-encoding genes that affect downstream fiber elongation. *GhMYB109* has minor or no role in regulating fiber initiation (Pu et al. 2008). A R3-MYB gene *GhCPC*, an ortholog of *TRY* in *Arabidopsis*, negatively regulates fiber initiation and early elongation (Liu et al. 2015). These early studies did not confirm the critical role of MYB genes in cotton fiber initiation with high degree of confidence.

The roles of cotton MYB genes *GhMYB25* and *GhMYB25-like* in fiber initiation were first studied by an Australian research group. Both MYB genes were identified by comparing the cDNA microarray profiles of 0 DPA ovules between a wild-type cotton and five fiber mutants (Wu et al. 2007; Wu et al. 2006). The expression of the *GhMYB25* promoter-GUS construct in the transgenic cotton was seen in the developing fibers and in the trichomes of a number of tissues including leaves, stems, etc. (Machado et al. 2009), which indicates that *GhMYB25* regulates specialized outgrowth of epidermal cells, including but not limited to cotton fibers. *GhMYB25-like* encodes a protein with 69% sequence identity to *GhMYB25*. RNAi-mediated gene silencing in transgenic cotton plants showed that *GhMYB25-like* is fiber specific and plays key role in the very early stage of fiber cell differentiation (Walford et al. 2011). *GhMYB25-like* functions upstream from other MYBs including *GhMYB25* and *GhMYB109* (Bedon et al. 2014).

There are numerous naturally occurring and man-made fiber mutants that display abnormal phenotypes including fuzzless, fiberless, short fiber, and immature fiber (Endrizzi and Ramsay 1980; Kohel 1972; Kohel et al. 1992; Naoumkina et al. 2017; Rong et al. 2005; Zhang and Pan 1991). Recently, the causative genes of three mutations affecting fiber initiation were identified, and all revealed as MYBMIXTA-like (MML) genes. MML family members contain the signature protein motif AQWESARxxAExRLxRES.

Fig. 8.1 Cotton fiber mutants XZ142 *fl*, N_1, n_2 and wild type

N_1 N_1 mutation results in fuzzless but linted seeds (Fig. 8.1). This locus was previously mapped on chromosome 12 (A12) (Endrizzi and Ramsay 1980; Kohel 1972). Wan et al. (2016) mapped the N_1 locus in a 49-kb region which contains only two MML genes: *GhMML3_A12* and *GhMML4_A12*. Further genetic analysis concluded *GhMML3_A12* (*Gh_A12G1503*) as the N_1 gene. This is the same gene as earlier-reported *GhMYB25-like* (Bedon et al. 2014; Walford et al. 2011). VIGS of *GhMML3_A12* in cotton resulted in fuzzless or reduced fuzz phenotype but not a fiberless phenotype as seen in transgenic cotton reported by Walford et al. (2011). Thus, Wan et al. (2016) concluded that *GhMML3_A12* regulates fuzz rather than lint fiber initiation.

n_2 Like N_1 mutation, n_2 causes fuzzless but linted seeds (Fig. 8.1). The n_2 gene resides on chromosome D12 that is homoeologous to chromosome A12. Zhu et al. (2018) proposed that the *GhMML3_D12* (*Gh_D12G1628*) is the best candidate for n_2 gene present in *G. barbadense* and the fiberless mutant XZ142 *fl*.

li_3 Zhang and Pan (1991) stated that fiber development in XZ142 *fl* (Fig. 8.1) is controlled by two pairs of recessive genes, n_2 responsible for fuzz fiber and li_3 for lint fiber initiation. Using populations derived from crosses between a n_2 line (lint and fuzzless, $n_2n_2Li_3Li_3$) and XZ142 *fl* (fiberless, $n_2n_2li_3li_3$), Wu et al. (2018) mapped the li_3 locus in a 79.8 kb region on chromosome D12. There are three genes, *GhMML3*, *GhMML4*, and a NAC protein, in this genomic interval. Sequence comparison between the two parents indicated that *GhMML4_D12* is the li_3 gene which suppresses lint fiber initiation in XZ142 *fl*. *GhMML4* was specifically expressed on chromosome D12 and not expressed or expressed at very low level on chromosome A12 during fiber initiation despite *GhMML4_A12* having normal genomic sequence. *GhMML4_A12* might have undergone de-functionalization after the formation of allotetraploid Upland cotton. The role of the NAC protein (*Gh_D12G2767*) is not yet clear.

It is clear that cotton fiber (lint and fuzz) initiation is controlled by MML genes especially *MML3* and *MML4*. However, why do cotton lines containing N_1 or n_2 gene have variable level of lint (and occasionally fuzz) density? Zhu et al. (2018) proposed that lint initiation is determined by the total expression level of both homeologs of *GhMML3* at ~0 DPA, while fuzz initiation is mainly governed by the expression level of *GhMML3_D12* at ~3 DPA. The fiberless seed phenotype in transgenic cotton plants as seen in Walford et al.'s (2011) experiment might be due to the silencing of both *GhMML3* homeologs in cotton genome.

There are ten pairs of MML genes in Upland cotton genome (Bedon et al. 2014; Zhang et al. 2015b). Three MML genes, i.e., *MML3 (MYB25-like)*, *MML4*, and *MML7 (MYB25)*, clearly have roles to regulate fiber initiation. The roles of other seven MML genes in fiber development remain to be investigated especially the three MMLs (8, 9, 10) residing on the chromosome pair A13/D13.

8.2.1.2 Other Trichome-Related Genes

Besides MYB genes, genes orthologous to *GL3*, *TTG1*, or homeodomain protein (GL2) also exist in cotton. Both *GhDEL65* and *GhDEL61*, bHLH family proteins in cotton, are functional homeologs of *AtGl3* or A*tEGl3* (Shangguan et al. 2016). HD1 is a homeodomain-leucine zipper transcription factor and plays important roles in seed and stem trichome development. Silencing of both homeologs of *GhHD1* by RNAi reduced trichome formation and delayed fiber initiation, while overexpression increased fiber initials but did not affect trichomes elsewhere (Walford et al. 2012). *GhTTG1* and *GhTTG3* were able to restore trichome formation in ttg1 mutant *Arabidopsis* plants (Humphries et al. 2005).

8.2.1.3 Additional Genes or Factors Affecting Fiber Initiation and Early Development

As early as in 1990, Hendrix (1990) found that the sucrose synthase (*SuSy*) gene plays an important role in young cotton ovule carbohydrate partitioning and development. Later Ruan and Chourey (1998) showed that fiber cell initiation in ovule epidermis is related to sucrose synthase expression alteration and carbon partitioning in developing seeds. Ruan (2005) suggested that *SuSy* affects fiber initiation partially by regulating cellulose precursor supply and turgor pressure within the fiber cell. A vacuolar invertase, *GhVIN1*, affects cotton fiber initiation, probably by regulating the transcription of several MYB genes and auxin signaling components required for fiber initiation (Wang et al. 2014). Likewise, jasmonic acid gene *GhJAZ2* functions as a primary transcription repressor by interacting with MYB genes to affect fiber initiation and elongation (Hu et al. 2016).

Early in vitro tissue culture experiments revealed important roles of plant hormones in fiber initiation and early development (Beasley and Ting 1973; Beasley and Ting 1974; Liao et al. 2010). Manipulation of auxin biosynthesis by overexpression of the IAA biosynthetic gene *GhiaaM* in cotton ovules greatly enhanced fiber initiation and eventual yield (Zhang et al. 2011). Overexpressing *GhGA20ox1*, a gene responsible for gibberellic acid synthesis, promoted fiber initiation and elongation (Xiao et al. 2010).

8.2.2 Genes Related to Fiber Elongation

After initiation, fiber cells enter a period of rapid elongation with peak around 8–12 DPA (Hinchliffe et al. 2011b; Naoumkina et al. 2015; Qin et al. 2007). Cotton fiber elongation is well studied, and many genes involved in this stage have been characterized (Table 8.1). Our current knowledge indicates that many genes are involved in the fiber elongation process. Of them, three major classes of genes are known to affect fiber elongation. First, cytoskeleton genes such as actins and tubulins are critical to fiber cell expansion. Second, to maintain high turgor pressure within a fiber cell is important for a sustainable cell elongation. Genes responsible for the biological processes such as the transportation of osmoticum are highly expressed during the elongation stage. Third, cell wall loosening is required for an efficient cell elongation. Genes involved in cell wall loosening directly or indirectly are upregulated at elongation stage. Plant hormones such as ethylene, auxins, and gibberellic acids are able to regulate the above genes and biological processes and consequently affect fiber elongation.

8.2.2.1 Actins, Tubulins, and Annexins

In the early 1990s, F-actin was shown to regulate microtubule orientation during cotton fiber development in an in vitro cytoskeletal disruption study (Seagull 1990). Later, Li et al. (2005) clearly demonstrated the role of an actin gene *GhACT1* in fiber elongation. Overexpression or downregulation of genes affecting actin binding (*GhPFN2*) (Wang et al. 2010b), actin depolymerizing (*GhADF1*) (Wang et al. 2009), or rearrangement of actin higher structures (*GhFIM2*) (Zhang et al. 2017a) resulted in fiber length changes in transgenic cottons. *GhCFE1A* functions as a dynamic linker between the actin cytoskeleton and the cortical endoplasmic reticulum network and plays an indirect role in fiber elongation (Lv et al. 2015). Overexpression of *GhCFE1A* in cotton not only delayed fiber cell elongation but also reduced the density of lint and fuzz fiber initials and stem trichomes.

The Ligon-lintless-1 (*Li*$_1$) mutant of cotton was discovered in 1929 and has been characterized as a single dominant gene located on chromosome D04 (chromosome 22) (Gilbert et al. 2013; Rong et al. 2005; Thyssen et al. 2015). Recently, Thyssen et al. (2017) reported that the causative gene of the *Li*$_1$ mutation is a substitution of glycine to valine at position 65 in the protein sequence of an actin gene, *GhACT_LI1* (*Gh_D04G0865*). This alteration disrupts progressive elongation of F-actin, resulting in a disorganized cytoskeleton and reduced cell polarity and consequently dwarf cotton plants with very short fibers.

Besides actins, tubulins, and annexins were found to affect fiber elongation. The gene *GhBTubL* can induce longitudinal growth of fission yeast (Ji et al. 2002). *GhTUA9*, an alpha-tubulin gene, was suggested to play important role in fiber elongation (Li et al. 2007b). RNAi-mediated downregulation of *GhAnn2*, an annexin gene, inhibited fiber elongation and secondary cell wall synthesis, resulting in shorter and thinner mature fibers in the transgenic plants (Tang et al. 2014a). Overexpression of *GhAnn3* increased leaf trichome density and length in transgenic *Arabidopsis* (Li et al. 2013a). The annexin genes located on chromosomes A10 and D10 in Upland cotton were thoroughly studied by Zhang et al. (2016). The A genome annexin gene *GhFAnnexA* affected cotton fiber elongation and was also involved in SCW biosynthesis. This gene may act as a Ca^{2+} conductance regulator and that reactive oxygen species (ROS) produced by *Rbohs* in a Ca^{2+}-dependent manner may determine fiber elongation caused by elevated intracellular turgor and cell wall loosening. The central role of *GhFAnnexA* may lie on its regulatory function of ROS production. ROS have been suggested to be involved in cell elongation by loosening the cell wall in coleoptiles, leaves, and roots (Liszkay et al. 2004).

8.2.2.2 Genes Affecting Cell Turgor Pressure

The turgor in a plant cell is achieved mainly through the influx of water driven by a relatively high concentration of osmoticum within a cell (Cosgrove 1997). To have this happen, a large number of plasma membrane and tonoplast intrinsic proteins (PIP and TIP) are required to express in the elongating fiber cells. Ruan (2005) previously had an excellent review in this subject. Recently, more experimental evidences support that maintaining a high turgor pressure is important for fiber elongation and many genes are involved in this process (Li et al. 2010a; Rungis et al. 2005; Tang et al. 2014a; Wang et al. 2010c, 2014). Zhang et al. (2017d) found that *GhSCP2D*, a putative sterol carrier protein gene, affects the close of plasmodesmata. Suppression of *GhSCP2D* caused earlier close of plasmodesmata, and upregulated a cohort of Sucrose Transport Proteins (SUT) and Sugars Will Eventually be Exported Transporters (SWEET) sucrose transport genes, and consequently affected fiber elongation.

Using the Ligon-lintless-1 and Ligon-lintless-2 short fiber mutants, Naoumkina et al. (2015) detected significantly lower osmotic pressure in the mutants than in the wild type. The major intrinsic proteins or aquaporins were the most overrepresented gene family among downregulated genes in both short fiber mutants. It has been shown that the osmotic water permeability is controlled by the activity of aquaporins (Javot et al. 2003; Peret et al. 2012; Postaire et al. 2010). Knockdown of expression of *GhPIP2* genes by RNA interference in *G. hirsutum* markedly inhibited fiber elongation (Li et al. 2013b). Thus, the reduced expression of aquaporins in short fiber mutants may reduce the influx of water into fiber cells and slow down the elongation process.

8.2.2.3 Cell Wall Loosening Genes

Ruan et al. (2001) demonstrated that a cell wall loosening gene expansin, *GhEXP1*, is critical to fiber elongation by unlocking the cross-linking of wall polysaccharides to allow the turgor-driven cell expansion during the active growth period. In *G. barbadense*, Li et al. (2016) identified two expansin genes, *GbEXPA2* and *GbEXPATR*. Overexpression of *GbEXPATR* resulted in longer, finer, and stronger fibers coupled with significantly thinner cell walls.

Besides expansins, another cell wall-loosening gene, xyloglucan endotransglycosylase/hydrolase (XTH), was upregulated in elongating fiber cells when cDNA profiles prepared from 10 DPA fibers of both WT cotton and fiberless mutant XZ142 *fl* were compared (Ji et al. 2003). In growing cells, xyloglucan is thought to connect cellulose microfibrils and regulate their separation during wall extension. Enzymes such as XTH that cleave and reattach xyloglucan polymers have significant effects on fiber elongation. Transgenic overexpression of *GhXTH1* in cotton increased fiber length up to 20% (Lee et al. 2010).

ROS is abundant in the actively elongating fiber cells (Qin and Zhu 2011). Their major function is suggested as cell wall loosening agents (Zhang et al. 2015a). Genes regulating ROS production such as *GhPOX1*, *GhAPX1*, *GhCDPK1*, and superoxide dismutase (*GhCSDs*) were found to affect fiber elongation (Kim et al. 2008; Kim and Triplett 2008; Li et al. 2007a; Mei et al. 2009; Zhang et al. 2015a). Tang et al. (2014b) had a thorough study on a gene encoding a calcium sensor, *GhCaM7*. Overexpressing *GhCaM7* promoted early fiber elongation, whereas *GhCaM7* suppression by RNAi delayed fiber initiation and inhibited fiber elongation. *GhCaM7* might modulate ROS production and act as a molecular link between Ca^{2+} and ROS signal pathways in fiber development. A well-characterized cotton gene, *GhWLIM1a*, has multiple functions to regulate fiber elongation and SCW biosynthesis. One of the functions is to mediate a cross talk between the cytoplasm and nucleus in developing cotton fibers in response to the ROS burst (Han et al. 2013).

8.2.2.4 Other Genes Related to Fiber Elongation

The role of ethylene in cotton fiber elongation was reported by Shi et al. (2006). The three 1-aminocyclopropane-1-carboxylic acid oxidase genes (*GhACO1-3*) responsible for ethylene production were expressed at significantly higher levels during this growth stage. Exogenously applied ethylene promoted fiber cell expansion, whereas fiber growth was suppressed by applying ethylene biosynthetic inhibitor L-(2-aminoethoxyvinyl)-glycine. They also indicated that ethylene may promote cell elongation by increasing the expression of sucrose synthase, tubulin, and expansin genes. Ethylene biosynthesis can be activated by saturated very-long-chain fatty acids (Qin et al. 2007).

GhHOX3 is an ortholog of *GL2* that is part of *Arabidopsis* leaf trichome protein complex. Transgenic analysis of *GhHOX3* showed that it plays prominent role in fiber elongation with very minor role in initiation. Silencing of *GhHOX3* reduced fiber length by more than 20%, whereas its overexpression led to longer fiber (Shan et al. 2014).

Arabinogalactan proteins (AGPs) are involved in many aspects of plant development. Overexpression of fasciclin-like arabinogalactan protein genes (*GhFLA1*, *GhAGP2/4*) promoted fiber elongation, leading to an increase of fiber length. *GhFLA1* and *GhAGP2/4* may function in fiber initiation and elongation by affecting AGP composition and the integrity of the primary cell wall matrix (Huang et al. 2013a; Li et al. 2010b). The gene *GhPRP5*, encoding a proline-rich protein, was found to affect fiber elongation (Xu et al. 2013b). The mechanism might be related to the production of hydroxyproline-rich glycoproteins that are cell wall components.

Cotton fiber elongation requires constant supply of energy. Recently, He et al. (2018) showed that two editing sites in *GhATP1* had a prominent difference in editing efficiency between fiber and ovule. These two vital editing sites are crucial for ATP synthase to produce sufficient ATP for cotton fiber cell elongation.

The gene Paclobutrazol Resistance 1 (*PRE1*) that encodes a basic helix-loop-helix (bHLH) transcription factor specific to cotton fiber cells is upregulated during rapid fiber elongation stage. Transgenic studies demonstrated that *PRE1* is a positive regulator of fiber elongation (Zhao et al. 2018).

8.2.3 Genes Related to SCW Thickening

Around 16–20 DPA, cotton fiber development transitions from elongation to SCW thickening. At SCW thickening stage, fiber cell wall becomes more rigid, and rapid cellulose synthesis occurs. Production of large amount of cellulose requires energy and carbon precursors dedicated to this process.

Ruan and co-authors (Ruan 2007; Ruan and Chourey 1998; Ruan et al. 2001, 2005) thoroughly studied the roles of sucrose synthase in cotton fiber development. Not only sucrose synthase involves in fiber initiation and elongation, it also is a partner in secondary wall cellulose biogenesis. Sucrose synthase cleaves sucrose to release fructose and UDP-glucose that is a substrate of cellulose (Brill et al. 2011). Overexpression of a sucrose synthase gene, *GhSusA1*, increased fiber length and strength, with the latter indicated by the enhanced thickening of cell wall during secondary wall formation stage (Jiang et al. 2012).

Cellulose is synthesized by a large membrane-bound protein complex that is named as cellulose synthesis complex comprised of multiple cellulose synthase proteins (CESAs) and other integral proteins (Doblin et al. 2002). The first cotton *CESAs* (a full-length *GhCesA1* and a 5'-truncated partial *GhCesA2*) were identified from developing cotton fibers during the SCW stage (Pear et al. 1996). According to the recent "hexamer of trimers" model, a cellulose synthesis complex is composed of 18 CESA proteins with 3 CESAs per lobe as the mostly likely composition of a rosette to account for an 18-chain cellulose microfibrils (Gonneau et al. 2014; Hill et al. 2014; Newman et al. 2013; Vandavasi et al. 2016). Upland cotton TM-1 genome contains at least 32 *GhCESA* family genes (Zhang et al. 2015b). They are classified into two major groups: one is expressed during primary cell wall development, and the other including at least 10 *GhCESA*s is predominately expressed during SCW biosynthesis. Tuttle et al. (2015) identified stage-specific cellulose synthases such as *GhCesA4*, *GhCesA7*, and *GhCesA8* at the SCW stage. In addition, a *chitinase-like* (Zhang et al. 2004), *WLIM1a* (Han et al. 2013), *GhADF1* (Wang et al. 2009), and *GhKNL1* (Gong et al. 2014) were reported to be involved in the SCW biosynthesis process of cotton fibers. Yadav et al. (2017) indicated that the *GhMYB1* regulates the expression of a lipase/hydrolase gene (*GhGDSL*) promoter to affect the fiber SCW thickening.

Recently, Yan et al. (2018) demonstrated that *GhTT2-3A* not only regulates pigment formation in brown cotton but also participates at the secondary wall thickening stage. This gene was previously identified by Hinchliffe et al. (2016) as linked to the brown color and fire retardancy in cotton. Zhang et al. (2018) found that a cotton NAC transcription factor (*GhFSN1*) is specifically expressed in SCW thickening stage. Upregulation of *GhFSN1* expression in cotton resulted in an increase in SCW thickness of fibers but a decrease in fiber length. Transcriptomic analysis revealed that *GhFSN1* activates or represses numerous downstream genes including *GhDUF231L1*, *GhKNL1*, *GhMYBL1*, *GhGUT1*, and *GhIRX12* genes.

The immature mutant (*im*) was first reported in 1970s (Kohel et al. 1974). Its fiber has thinner cell wall than wild-type cotton (Kim et al. 2013). Thyssen et al. (2016) identified a pentatricopeptide repeat (PPR) gene (*Gh_A03G0489*) as the causative gene of the *im* locus. The *im* mutant has a 22 bp deletion in the coding sequence of the PPR gene. The 22 bp deletion in the *im* mutant should abolish the function of the PPR gene *Gh_A03G0489*. Analysis of *Gh_A03G0489* indicates that it contains a mitochondria-specific transit peptide and 25 PPR repeats with a specific predicted RNA-binding motif. Transcriptome analysis work by Kim et al. (2013) strongly implicated mitochondrial function in the development of the immature fiber mutant phenotype. It is speculated that energy distribution via the function of mitochondria is important to the SCW thickening.

8.2.4 Other Genes Affecting Fiber Development

Although not specifically to fiber, the cotton *PHYA1* gene greatly affects cotton plant mass and fiber yield and quality. Suppression of the *PHYA1* transcript and compensatory overexpression of the remaining phytochromes exhibited vigorous root and vegetative growth, early-flowering, significantly improved fiber length and other fiber characteristics (Abdurakhmonov et al. 2014).

8.3 Fiber Quantitative Trait Loci

Cotton fiber quality attributes are complex traits that are controlled by multiple genes and greatly affected by environmental factors (Lacape et al. 2010; Paterson et al. 2003; Wang et al. 2015). Since the first cotton genetic map was published in 1994 (Reinisch et al. 1994), cotton scientists have been developing molecular markers associated with fiber quality QTL. So far, thousands of fiber QTL have been identified (Chen et al. 2009; Du et al. 2018; Fang 2015; Fang et al. 2014; Kumar et al. 2012; Lacape et al. 2005, 2010; Ma et al. 2017, 2018; Mei et al. 2004; Paterson et al. 2003; Qin et al. 2008; Said et al. 2013; Shen et al. 2011; Sun et al. 2017; Wang et al. 2006, 2007, 2012; Yu et al. 2010; Zhang et al. 2012, 2013, 2017c). Previously, Said et al. (2013) conducted a meta-QTL analysis for fiber quality based on 42 publications. They listed 810 fiber QTL and their locations. In addition, they also identified 76 fiber QTL clusters or hot spots. In this review, I will focus on the studies published since 2013 and more specifically on the relatively stable fiber length and strength QTL that were identified by multiple research groups and can be used in breeding through MAS.

8.3.1 Methods for Fiber QTL Identification

A population of plants that demonstrate fiber quality differences is required for fiber QTL identification. These populations include those derived from a cross between two parents or crosses from multiple parents. Recently, GWAS using natural varieties has attracted great attention in fiber QTL identification.

QTL Identification Using a Biparental Population This has been the most used method to identify fiber QTL in cotton. Almost all fiber QTL studies published before 2013 were based on the analysis of progeny plants derived from a cross between two parents. Readers are suggested to refer the publication (Fang 2015) for detail QTL analysis method using a biparental population.

QTL Identification Using a Multiple Parent Population As indicated by its name, more than two parents are used to develop a population for QTL identification. Due to very low genetic diversity within Upland cotton, it is very difficult or impossible to obtain a high-resolution genetic map if using a biparental population. The approach using populations involving more than two parents ensured greater genetic diversity and an increased polymorphism frequency in the mapping populations and improved possibility of QTL analysis. In cotton, populations derived from 3 (Zhang et al. 2012), 4 (Qin et al. 2008), 8 (Huang et al. 2018), 10 (Zeng et al. 2009), and 11 parents (Fang et al. 2014; Islam et al. 2016b) were used for fiber QTL identification.

Genome- Wide Association Mapping Association mapping based on linkage disequilibrium has become a powerful tool to identify marker-trait QTL associations in plants (Hamblin et al. 2011). It utilizes all alleles containing in a panel of varieties and phenotypic data that may be obtained through experiments or historical variety tests. In cotton, a few association mapping reports were published, and most focused on fiber or yield traits (Abdurakhmonov et al. 2009; Ademe et al. 2017; Cai et al. 2014; Du et al. 2018; Huang et al. 2017; Ma et al. 2018; Mei et al. 2013; Sun et al. 2017; Zhang et al. 2013). In addition, the US cotton community is developing a nested association mapping (NAM) population for more rigorous trait QTL identification.

8.3.2 Length QTL

Fiber length (FL) is largely influenced by the genetic background of each cultivar. The inheritability of fiber length is quite high although adverse environmental conditions could reduce FL below a genotype's potential (Bradow and Davidonis 2000; Meredith et al. 2012). Recently a few relatively stable FL QTL were identified by multiple researchers using different populations or different mapping strategies. The first one is the QTL on the chromosome D11 (Chr. 21). Islam et al. (2016b) identified a significant FL QTL at 26 Mb region of Chr. D11 based on the GWAS of an Upland cotton MAGIC population. This QTL was confirmed by Sun et al. (2017) and Huang et al. (2017) based on GWAS of cotton varieties, most of them Chinese varieties. The second one is the QTL on Chr. A01. Previously, Shen et al. (2011) identified this FL QTL. This QTL was introgressed from *G. barbadense* into Upland cotton and responsible for 24% of the phenotypic variation. Xu et al. (2017) further fine-tuned the QTL region and narrowed it down to a 0.9 cM region of Chr. A01. DNA markers flanking this QTL were developed and can be used via MAS to transfer this QTL into other elite cotton lines.

8.3.3 Strength QTL

Many fiber strength (FS) QTL have been identified, and among them two are quite stable. The first is the QTL on chromosome A07. Using a MAGIC population, the author (Fang et al. 2014) and his colleagues identified this FS QTL. Islam et al. (2016b) further enriched this QTL region with GBS-based SNP markers. The QTL was mapped around 72–76 Mb region of Chr. A07. Hugie et al. (2016) tested the transferability of this FS QTL in different genetic backgrounds. This same QTL was also identified by four different research groups using *Gossypium barbadense* introgressed lines (Cao et al. 2015), biparental population mapping (Fang et al. 2017b; Zhang et al. 2017c), and GWAS of cotton varieties (Huang et al. 2017; Sun et al. 2017; Ma et al. 2018). This QTL has significant effect on fiber strength. Survey of this QTL in a panel of US cotton varieties indicated that this FS QTL might originate from *G. barbadense* (Islam et al. 2016b). Interestingly, at this FS QTL region, QTL for fiber length, short fiber content, uniformity, and fiber maturity are also present (Cao et al. 2015; Islam et al. 2016b). This FS QTL is being used in practical breeding in the USA to improve fiber strength. The second is the FS QTL on Chr. 24 (D08). This QTL was first identified in the Chinese germplasm line Suyuan 7235 (Kumar et al. 2012). The favorable allele from Suyuan 7235 explained up to 40% of the total phenotypic variation and accounted for an increase of up to 22.8 kN mkg^{-1}. Kumar et al. (2012) developed DNA markers for this QTL which were being used in MAS for FS improvement. Islam et al. (2016c) identified a significant FS QTL on Chr. 24 that is present in the cotton line MD52ne. Comparison of the markers associated to these QTL indicates that these two FS QTL on Chr. 24 are possibly the same.

8.3.4 MIC QTL

MIC is a measurement of the air permeability through a mass of fiber compressed to a fixed volume and is influenced by both fiber fineness and maturity (Montalvo Jr. 2005). Unlike FL and FS that are simple traits from a biological point of view, MIC is a complex trait that consists of both fiber maturity and fineness and is greatly affected by environmental factors (Paterson et al. 2003). Moreover, interpretation of the MIC is complex as well. A given MIC value may be reached by very different combinations of fineness and maturity (Lacape et al. 2010). Nevertheless, a MIC reading is still widely used in both the breeding and marketing of cotton. Since MIC is a complex trait, it has been difficult to find MIC QTL that are consistent or congruent among research publications. In Said et al.'s (2013) research, they found 20 MIC hot spots, the most among all fiber traits. In spite of difficulty to identify stable MIC QTL, Said et al. (2013) indicated that Chr. 24 contains three MIC QTL hot spots. Recently, the author (unpublished data) identified a significant MIC QTL on Chr. 24 using an Upland MAGIC population. How stable this MIC QTL will be in other genetic background remains to be investigated.

8.4 Conclusions and Future Perspectives

During the past two decades, a lot of accomplishments and progresses have been made to identify genes regulating cotton fiber development. Thousands of genes were found to be related to fiber development directly or indirectly, and many have been characterized through transgenic analysis. The causative genes of five fiber mutations were identified via genetic mapping. MML transcription factors play an important role in cotton fiber initiation. The expression levels of MML3 homeologs at specific times of fiber development determine lint and fuzz fiber initiation. Cytoskeleton actins, tubulins, and annexins are essential for fiber elongation. In order to enable a fiber cell to have a sustainable elongation, its cell wall needs to be loosed, and a high turgor pressure within its cell needs to be maintained. At the SCW thickening stage, a fiber cell becomes a bio-factory for cellulose production with more rigid cell wall. A sustainable supply of energy and carbon precursors are important for cellulose synthesis. Plant hormones such as ethylene, auxins, and gibberellic acids affect all stages of fiber development by exerting regulatory influences on the genes involved.

In the aspect of fiber QTL identification, cotton researchers have identified thousands of QTL affecting fiber properties using a variety of mapping strategies. Many fiber QTL clusters and hot spots have been identified. In addition, a few stable fiber QTL were repeatedly identified by different research groups using different methods. Some of these QTL are being used in practical breeding.

In spite of these accomplishments, many fundamental questions and practical challenges remain to be addressed. Although many genes were characterized via transgenic analysis, only one or two have been tested in large-scale field planting. So far, there is not a single commercial variety that results from the transformation of a fiber gene. The utilization of MAS to improve fiber quality in practical breeding is still at its infant stage. In order to translate the gained knowledge into practical utilization, I think that future research should be in the following areas:

1. More rigorous and thorough analysis of transgenic cotton plants in a large-scale field testing for both agronomic and fiber traits. Although many fiber genes were transformed into cotton, their overall field performance was rarely reported. The gap between a laboratory analysis and a field test needs to be narrowed in order for us to reap the benefit of a fiber-improving gene.
2. Functional characterization of important fiber genes by silencing and overexpression in transgenic cotton. Due to difficulties of cotton transformation, most of the identified fiber mutant genes, i.e., n_2, li_3, Li_1, and *im*, have not been transformed into cotton to further explore their functions in fiber development. RNAi or antisense silencing sometimes may unintentionally target both homeologs as seen in Walford et al.'s experiment (2011) and other members of a gene family. Therefore another technique such as CRISPR/Cas9 should be considered in knockdown (or alteration) of specific homoeologous genes.
3. As cotton fiber development involves many genes, it is important to understand the gene network and how the genes cross-talk to regulate fiber development. To

elucidate a gene network is an obviously difficult long-term research task. However, without such knowledge, it will be difficult to improve fiber quality and yield simultaneously through a biotechnological approach.

4. Roles of plant hormones in fiber development. Future researches should understand how plant hormones affect fiber development in following aspects: (a) interaction between plant hormone genes and fiber genes, (b) measurement of endogenous levels of plant hormones and their correlation with fiber development, and (c) effects of exogenous application of plant hormones on fiber development.
5. Confirmation and utilization of stable fiber QTL in practical breeding via MAS. Fiber QTL with large effect (>2% in explaining phenotypic variance) should be validated in different genetic backgrounds and in different environments. If validated, this fiber QTL and related DNA markers should be transferred to cotton breeders for use in breeding.
6. Micronaire is a complex trait that includes both maturity and fineness. Although many MIC QTL were identified, very few were repeatedly identified or confirmed in different experiments by different research groups. It is recommended to dissect a particular important MIC QTL into maturity and fineness and further identify maturity and fineness QTL separately. This will require a more accurate measurement of fiber maturity and fineness using a Cottonscope or other specific instrument.
7. Identification of fiber genes based on in-depth analysis of a fiber QTL. Usually a fiber QTL region spans several Mb of genomic interval. There may be many genes within a QTL. Some of them may not be involved in fiber development. How genes contained in the same fiber QTL regulate fiber development has not been studied. Cotton researchers should dissect a stable fiber QTL (such as strength or length QTL) to identify specific fiber genes. Identification of fiber genes at the QTL site will enable precise introgression of a target gene(s) in breeding.

Acknowledgment I thank Dr. Tianzhen Zhang in Zhejiang University, China, for providing the pictures of N_1 and n_2 mutants. The author's research is funded by the USDA-ARS project #6054-21000-017-00D. Mention of trade names or commercial products in this article is solely for the purpose of providing specific information and does not imply recommendation or endorsement by the US Department of Agriculture that is an equal opportunity provider and employer.

References

Abdurakhmonov IY, Buriev ZT, Saha S, Jenkins JN, Abdukarimov A, Pepper AE (2014) Phytochrome RNAi enhances major fibre quality and agronomic traits of the cotton Gossypium hirsutum L. Nat Commun 5:3062

Abdurakhmonov IY, Saha S, Jenkins JN, Buriev ZT, Shermatov SE, Scheffler BE, Pepper AE, Yu JZ et al (2009) Linkage disequilibrium based association mapping of fiber quality traits in G. hirsutum L. variety germplasm. Genetica 136:401–417

Ademe MS, He S, Pan Z, Sun J, Wang Q, Qin H, Liu J, Liu H et al (2017) Association mapping analysis of fiber yield and quality traits in upland cotton (Gossypium hirsutum L.). Mol Genet Genomics 292:1267–1280

Aleman L, Kitamura J, Abdel-mageed H, Lee J, Sun Y, Nakajima M, Ueguchi-Tanaka M, Matsuoka M et al (2008) Functional analysis of cotton orthologs of GA signal transduction factors GID1 and SLR1. Plant Mol Biol 68:1–16

Avci U, Pattathil S, Singh B, Brown VL, Hahn MG, Haigler CH (2013) Cotton fiber cell walls of Gossypium hirsutum and Gossypium barbadense have differences related to loosely-bound xyloglucan. PLoS One 8:e56315

Beasley CA, Ting IP (1973) The effects of plant growth substances on in vitro fiber development from fertilized cotton ovules. Am J Bot 60:130–139

Beasley CA, Ting IP (1974) Effects of plant growth substances on in vitro fiber development from unfertilized cotton ovules. Am J Bot 61:188–194

Bedon F, Ziolkowski L, Walford SA, Dennis ES, Llewellyn DJ (2014) Members of the MYBMIXTA-like transcription factors may orchestrate the initiation of fiber development in cotton seeds. Front Plant Sci 5:179

Bradow JM, Davidonis GH (2000) Quantitation of fiber quality and the cotton production-processing interface: a physiologist's perspective. J Cotton Sci 4:34–64

Brill E, van Thournout M, White RG, Llewellyn D, Campbell PM, Engelen S, Ruan YL, Arioli T et al (2011) A novel isoform of sucrose synthase is targeted to the cell wall during secondary cell wall synthesis in cotton fiber. Plant Physiol 157:40–54

Cai C, Ye W, Zhang T, Guo W (2014) Association analysis of fiber quality traits and exploration of elite alleles in upland cotton cultivars/accessions (Gossypium hirsutum L.). J Integr Plant Biol 56:51–62

Cao Z, Zhu X, Chen H, Zhang T (2015) Fine mapping of clustered quantitative trait loci for fiber quality on chromosome 7 using a Gossypium barbadense introgressed line. Mol Breed 35:215

Chen H, Qian N, Guo W, Song Q, Li B, Deng F, Dong C, Zhang T (2009) Using three overlapped RILs to dissect genetically clustered QTL for fiber strength on Chro.D8 in upland cotton. Theor Appl Genet 119:605–612

Cosgrove DJ (1997) Relaxation in a high-stress environment: the molecular bases of extensible cell walls and cell enlargement. Plant Cell 9:1031–1041

Deng F, Tu L, Tan J, Li Y, Nie Y, Zhang X (2012) GbPDF1 is involved in cotton fiber initiation via the core cis-element HDZIP2ATATHB2. Plant Physiol 158:890–904

Ding M, Jiang Y, Cao Y, Lin L, He S, Zhou W, Rong J (2014) Gene expression profile analysis of Ligon lintless-1 (Li1) mutant reveals important genes and pathways in cotton leaf and fiber development. Gene 535:273–285

Doblin MS, Kurek I, Jacob-Wilk D, Delmer DP (2002) Cellulose biosynthesis in plants: from genes to rosettes. Plant Cell Physiol 43:1407–1420

Du X, Huang G, He S, Yang Z, Sun G, Ma X, Li N, Zhang X et al (2018) Resequencing of 243 diploid cotton accessions based on an updated A genome identifies the genetic basis of key agronomic traits. Nat Genet 50:796–802

Endrizzi JE, Ramsay G (1980) Identification of ten chromosome deficiencies in cotton. J Hered 71:45–48

Fang DD (2015) Molecular breeding. In: Fang DD, Percy RG (eds) Cotton, 2nd edn. ASA-CSSA-SSSA, Madison, pp 255–288

Fang DD, Jenkins JN, Deng DD, McCarty JC, Li P, Wu J (2014) Quantitative trait loci analysis of fiber quality traits using a random-mated recombinant inbred population in upland cotton (Gossypium hirsutum L.). BMC Genomics 15:397

Fang L, Wang Q, Hu Y, Jia Y, Chen J, Liu B, Zhang Z, Guan X et al (2017a) Genomic analyses in cotton identify signatures of selection and loci associated with fiber quality and yield traits. Nat Genet 49:1089–1098

Fang X, Liu X, Wang X, Wang W, Liu D, Zhang J, Liu D, Teng Z et al (2017b) Fine-mapping qFS07.1 controlling fiber strength in upland cotton (Gossypium hirsutum L.). Theor Appl Genet 130:795–806

Gilbert MK, Turley RB, Kim HJ, Li P, Thyssen G, Tang Y, Delhom CD, Naoumkina M et al (2013) Transcript profiling by microarray and marker analysis of the short cotton (Gossypium hirsutum L.) fiber mutant Ligon lintless-1 (Li1). BMC Genomics 14:403

Gong SY, Huang GQ, Sun X, Qin LX, Li Y, Zhou L, Li XB (2014) Cotton KNL1, encoding a class II KNOX transcription factor, is involved in regulation of fibre development. J Exp Bot 65:4133–4147

Gonneau M, Desprez T, Guillot A, Vernhettes S, Hofte H (2014) Catalytic subunit stoichiometry within the cellulose synthase complex. Plant Physiol 166:1709–1712

Gore MA, Fang DD, Poland JA, Zhang J, Percy RG, Cantrell RG, Thyssen G, Lipka AE (2014) Linkage map construction and quantitative trait locus analysis of agronomic and fiber quality traits in cotton. Plant Genom 7:1–8

Guan X, Pang M, Nah G, Shi X, Ye W, Stelly DM, Chen ZJ (2014) miR828 and miR858 regulate homoeologous MYB2 gene functions in Arabidopsis trichome and cotton fibre development. Nat Commun 5:3050

Haigler CH, Betancur L, Stiff MR, Tuttle JR (2012) Cotton fiber: a powerful single-cell model for cell wall and cellulose research. Front Plant Sci 3:104

Haigler CH, Zhang D, Wilkerson CG (2005) Biotechnological improvement of cotton fibre maturity. Physiol Plant 124:285–294

Hamblin MT, Buckler ES, Jannink JL (2011) Population genetics of genomics-based crop improvement methods. Trends Genet 27:98–106

Han LB, Li YB, Wang HY, Wu XM, Li CL, Luo M, Wu SJ, Kong ZS et al (2013) The dual functions of WLIM1a in cell elongation and secondary wall formation in developing cotton fibers. Plant Cell 25:4421–4438

Hao J, Tu L, Hu H, Tan J, Deng F, Tang W, Nie Y, Zhang X (2012) GbTCP, a cotton TCP transcription factor, confers fibre elongation and root hair development by a complex regulating system. J Exp Bot 63:6267–6281

Harmer SE, Orford SJ, Timmis JN (2002) Characterisation of six alpha-expansin genes in Gossypium hirsutum (upland cotton). Mol Genet Genomics 268:1–9

He P, Xiao G, Liu H, Zhang L, Zhao L, Tang M, Huang S, An Y et al (2018) Two pivotal RNA editing sites in the mitochondrial atp1mRNA are required for ATP synthase to produce sufficient ATP for cotton fiber cell elongation. New Phytol 218:167–182

Hendrix DL (1990) Carbohydrates and carbohydrate enzymes in developing cotton ovules. Physiol Plant 78:85–92

Hill JL, Hammudi MB, Tien M (2014) The Arabidopsis cellulose synthase complex: a proposed hexamer of CESA trimers in an equimolar stoichiometry. Plant Cell 26:4834–4842

Hinchliffe DJ, Condon BD, Thyssen G, Naoumkina M, Madison CA, Reynolds M, Delhom CD, Fang DD et al (2016) The GhTT2_A07 gene is linked to the brown colour and natural flame retardancy phenotypes of Lc1 cotton (Gossypium hirsutum L.) fibres. J Exp Bot 67:5461–5471

Hinchliffe DJ, Meredith WR, Delhom CD, Thibodeaux DP, Fang DD (2011a) Elevated growing degree days influence transition stage timing during cotton (Gossypium hirsutum L.) fiber development and result in increased fiber strength. Crop Sci 51:1683–1692

Hinchliffe DJ, Meredith WR, Yeater KM, Kim HJ, Woodward AW, Chen ZJ, Triplett BA (2010) Near-isogenic cotton germplasm lines that differ in fiber-bundle strength have temporal differences in fiber gene expression patterns as revealed by comparative high-throughput profiling. Theor Appl Genet 120:1347–1366

Hinchliffe DJ, Turley RB, Naoumkina M, Kim HJ, Tang Y, Yeater KM, Li P, Fang DD (2011b) A combined functional and structural genomics approach identified an EST-SSR marker with complete linkage to the Ligon lintless-2 genetic locus in cotton (Gossypium hirsutum L.). BMC Genomics 12:445

Hu H, He X, Tu L, Zhu L, Zhu S, Ge Z, Zhang X (2016) GhJAZ2 negatively regulates cotton fiber initiation by interacting with the R2R3-MYB transcription factor GhMYB25-like. Plant J 88:921–935
Huang C, Nie X, Shen C, You C, Li W, Zhao W, Zhang X, Lin Z (2017) Population structure and genetic basis of the agronomic traits of upland cotton in China revealed by a genome-wide association study using high-density SNPs. Plant Biotechnol J 15:1374–1386
Huang C, Shen C, Wen T, Gao B, Zhu LX, Ahmed MM, Li D et al (2018) SSR-based association mapping of fiber quality in upland cotton using an eight-way MAGIC population. Mol Gen Genomics 293(4):793–805. https://doi.org/10.1007/s00438-00018-01419-00434
Huang GQ, Gong SY, Xu WL, Li W, Li P, Zhang CJ, Li DD, Zheng Y et al (2013a) A fasciclin-like arabinogalactan protein, GhFLA1, is involved in fiber initiation and elongation of cotton. Plant Physiol 161:1278–1290
Huang QS, Wang HY, Gao P, Wang GY, Xia GX (2008) Cloning and characterization of a calcium dependent protein kinase gene associated with cotton fiber development. Plant Cell Rep 27:1869–1875
Huang Y, Liu X, Tang K, Zuo K (2013b) Functional analysis of the seed coat-specific gene GbMYB2 from cotton. Plant Physiol Biochem 73:16–22
Huang Y, Wang J, Zhang L, Zuo K (2013c) A cotton annexin protein AnxGb6 regulates fiber elongation through its interaction with actin 1. PLoS One 8:e66160
Hugie KL, Fang DD, Smith WC, Li P, Hinze LL, HS S, Jones DC (2016) Utility assessment of published microsatellite markers for fiber length and bundle strength QTL in a cotton breeding program. Crop Sci 56:2983–2995
Hulskamp M (2004) Plant trichomes: a model for cell differentiation. Nat Rev Mol Cell Biol 5:471–480
Hulskamp M, Misra S, Jurgens G (1994) Genetic dissection of trichome cell development in Arabidopsis. Cell 76:555–566
Humphries JA, Walker AR, Timmis JN, Orford SJ (2005) Two WD-repeat genes from cotton are functional homologues of the Arabidopsis thaliana TRANSPARENT TESTA GLABRA1 (TTG1) gene. Plant Mol Biol 57:67–81
Islam MS, Fang DD, Thyssen GN, Delhom CD, Liu Y, Kim HJ (2016a) Comparative fiber property and transcriptome analyses reveal key genes potentially related to high fiber strength in cotton (Gossypium hirsutum L.) line MD52ne. BMC Plant Biol 16:36
Islam MS, Thyssen GN, Jenkins JN, Zeng L, Delhom CD, McCarty JC, Deng DD, Hinchliffe DJ et al (2016b) A MAGIC population-based genome-wide association study reveals functional association of GhRBB1_A07 gene with superior fiber quality in cotton. BMC Genomics 17:903
Islam MS, Zeng L, Thyssen GN, Delhom CD, Kim HJ, Li P, Fang DD (2016c) Mapping by sequencing in cotton (Gossypium hirsutum) line MD52ne identified candidate genes for fiber strength and its related quality attributes. Theor Appl Genet 129:1071–1086
Javot H, Lauvergeat V, Santoni V, Martin-Laurent F, Guclu J, Vinh J, Heyes J, Franck KI et al (2003) Role of a single aquaporin isoform in root water uptake. Plant Cell 15:509–522
Ji S, Lu Y, Li J, Wei G, Liang X, Zhu Y (2002) A beta-tubulin-like cDNA expressed specifically in elongating cotton fibers induces longitudinal growth of fission yeast. Biochem Biophys Res Commun 296:1245–1250
Ji SJ, Lu YC, Feng JX, Wei G, Li J, Shi YH, Fu Q, Liu D et al (2003) Isolation and analyses of genes preferentially expressed during early cotton fiber development by subtractive PCR and cDNA array. Nucleic Acids Res 31:2534–2543
Jiang Y, Guo W, Zhu H, Ruan YL, Zhang T (2012) Overexpression of GhSusA1 increases plant biomass and improves cotton fiber yield and quality. Plant Biotechnol J 10:301–312
Kim HJ, Kato N, Kim S, Triplett B (2008) Cu/Zn superoxide dismutases in developing cotton fibers: evidence for an extracellular form. Planta 228:281–292

Kim HJ, Tang Y, Moon HS, Delhom CD, Fang DD (2013) Functional analyses of cotton (Gossypium hirsutum L.) immature fiber (im) mutant infer that fiber cell wall development is associated with stress responses. BMC Genomics 14:889
Kim HJ, Triplett B (2008) Involvement of extracellular Cu/Zn superoxide dismutase in cotton fiber primary and secondary cell wall biosynthesis. Plant Signal Behav 3:1119–1121
Kim HJ, Triplett BA (2001) Cotton fiber growth in planta and in vitro. Models for plant cell elongation and cell wall biogenesis. Plant Physiol 127:1361–1366
Kim HJ, Triplett BA (2004) Characterization of GhRac1 GTPase expressed in developing cotton (Gossypium hirsutum L.) fibers. Biochim Biophys Acta 1679:214–221
Kim HJ, Triplett BA, Zhang HB, Lee MK, Hinchliffe DJ, Li P, Fang DD (2012) Cloning and characterization of homeologous cellulose synthase catalytic subunit 2 genes from allotetraploid cotton (Gossypium hirsutum L.). Gene 494:181–189
Kohel RJ (1972) Linkage tests in upland cotton, Gossypium hirsutum L. II. Crop Sci 12:66–69
Kohel RJ, Narbuth EV, Benedict CR (1992) Fiber development of Ligon lintless-2 mutant of cotton. Crop Sci 32:733–735
Kohel RJ, Quisenberry JE, Benedict CR (1974) Fiber elongation and dry weight changes in mutant lines of cotton. Crop Sci 14:471–474
Kumar P, Singh R, Lubbers EL, Shen X, Paterson AH, Campbell BT, Jones DC, Chee PW (2012) Mapping and validation of fiber strength quantitative trait loci on chromosome 24 in upland cotton. Crop Sci 52:1115–1122
Lacape JM, Llewellyn D, Jacobs J, Arioli T, Becker D, Calhoun S, Al-Ghazi Y, Liu S et al (2010) Meta-analysis of cotton fiber quality QTLs across diverse environments in a Gossypium hirsutum x G. barbadense RIL population. BMC Plant Biol 10:132
Lacape JM, Nguyen TB, Courtois B, Belot JL, Giband M, Gourlot JP, Gawryziak G, Roques S et al (2005) QTL analysis of cotton fiber quality using multiple G. hirsutum × G. barbadense backcross generations. Crop Sci 45:123–140
Larkin JC, Oppenheimer DG, Marks MD (1994) The GL1 gene and the trichome developmental pathway in Arabidopsis thaliana. Results Probl Cell Differ 20:259–275
Lee J, Burns TH, Light G, Sun Y, Fokar M, Kasukabe Y, Fujisawa K, Maekawa Y et al (2010) Xyloglucan endotransglycosylase/hydrolase genes in cotton and their role in fiber elongation. Planta 232:1191–1205
Lee JJ, Hassan OS, Gao W, Wei NE, Kohel RJ, Chen XY, Payton P, Sze SH et al (2006) Developmental and gene expression analyses of a cotton naked seed mutant. Planta 223:418–432
Lee JJ, Woodward AW, Chen ZJ (2007) Gene expression changes and early events in cotton fibre development. Ann Bot 100:1391–1401
Li B, Li DD, Zhang J, Xia H, Wang XL, Li Y, Li XB (2013a) Cotton AnnGh3 encoding an annexin protein is preferentially expressed in fibers and promotes initiation and elongation of leaf trichomes in transgenic Arabidopsis. J Integ Plant Biol 55:902–916
Li B, Yang Y, Hu WR, Li XD, Cao JQ, Fan L (2015) Over-expression of GhUGP1 in upland cotton improves fibre quality and reduces fibre sugar content. Plant Breed 134:197–202
Li DD, Ruan XM, Zhang J, Wu YJ, Wang XL, Li XB (2013b) Cotton plasma membrane intrinsic protein 2s (PIP2s) selectively interact to regulate their water channel activities and are required for fibre development. New Phytol 199:695–707
Li HB, Qin YM, Pang Y, Song WQ, Mei WQ, Zhu YX (2007a) A cotton ascorbate peroxidase is involved in hydrogen peroxide homeostasis during fibre cell development. New Phytol 175:462–471
Li L, Wang XL, Huang GQ, Li XB (2007b) Molecular characterization of cotton GhTUA9 gene specifically expressed in fibre and involved in cell elongation. J Exp Bot 58:3227–3238
Li XB, Fan XP, Wang XL, Cai L, Yang WC (2005) The cotton ACTIN1 gene is functionally expressed in fibers and participates in fiber elongation. Plant Cell 17:859–875
Li XR, Wang L, Ruan YL (2010a) Developmental and molecular physiological evidence for the role of phosphoenolpyruvate carboxylase in rapid cotton fibre elongation. J Exp Bot 61:287–295

Li Y, Liu D, Tu L, Zhang X, Wang L, Zhu L, Tan J, Deng F (2010b) Suppression of GhAGP4 gene expression repressed the initiation and elongation of cotton fiber. Plant Cell Rep 29:193–202
Li Y, Tu L, Pettolino FA, Ji S, Hao J, Yuan D, Deng F, Tan J et al (2016) GbEXPATR, a species-specific expansin, enhances cotton fibre elongation through cell wall restructuring. Plant Biotechnol J 14:951–963
Liao WB, Zhang J, Xua NF, Peng M (2010) The role of phytohormones in cotton fiber development. Russ J Plant Physiol 57:462–468
Liszkay A, van der Zalm E, Schopfer P (2004) Production of reactive oxygen intermediates ($O_2^{\cdot -}$, H_2O_2, and $^{\cdot}OH$) by maize roots and their role in wall loosening and elongation growth. Plant Physiol 136:3114–3123
Liu B, Zhu Y, Zhang T (2015) The R3-MYB gene GhCPC negatively regulates cotton fiber elongation. PLoS One 10:e0116272
Loguerico LL, Zhang JQ, Wilkins TA (1999) Differential regulation of six novel MYB-domain genes defines two distinct expression patterns in allotetraploid cotton (Gossypium hirsutum L.). Mol Gen Genet 261:660–671
Lv F, Wang H, Wang X, Han L, Ma Y, Wang S, Feng Z, Niu X et al (2015) GhCFE1A, a dynamic linker between the ER network and actin cytoskeleton, plays an important role in cotton fibre cell initiation and elongation. J Exp Bot 66:1877–1889
Ma L, Zhao Y, Wang Y, Shang L, Hua J (2017) QTLs analysis and validation for fiber quality traits using maternal backcross population in upland cotton. Front Plant Sci 8:2168
Ma Z, He S, Wang X, Sun J, Zhang Y, Zhang G, Wu L, Li Z et al (2018) Resequencing a core collection of upland cotton identifies genomic variation and loci influencing fiber quality and yield. Nat Genet 50:803–813
Machado A, Wu Y, Yang Y, Llewellyn DJ, Dennis ES (2009) The MYB transcription factor GhMYB25 regulates early fibre and trichome development. Plant J 59:52–62
Mei H, Zhu X, Zhang T (2013) Favorable QTL alleles for yield and its components identified by association mapping in Chinese upland cotton cultivars. PLoS One 8:e82193
Mei M, Syed NH, Gao W, Thaxton PM, Smith CW, Stelly DM, Chen ZJ (2004) Genetic mapping and QTL analysis of fiber-related traits in cotton (Gossypium). Theor Appl Genet 108:280–291
Mei W, Qin Y, Song W, Li J, Zhu Y (2009) Cotton GhPOX1 encoding plant class III peroxidase may be responsible for the high level of reactive oxygen species production that is related to cotton fiber elongation. J Genet Genomics 36:141–150
Meredith WR, Boykin DL, Bourland FM, Caldwell WD, Campbell T, Gannaway JR, Glass K, Jones AP et al (2012) Genotype X environment interactions over seven years for yield, yield components, fiber quality, and gossypol traits in the regional high quality tests. J Cotton Sci 16:160–169
Montalvo JG Jr (2005) Relationships between micronaire, fineness, and maturity. Part I. Fundamentals. J Cotton Sci 9:81–88
Naoumkina M, Bechere E, Fang DD, Thyssen GN, Florane CB (2017) Genome-wide analysis of gene expression of EMS-induced short fiber mutant Ligon lintless-y (liy) in cotton (Gossypium hirsutum L.). Genomics 109:320–329
Naoumkina M, Thyssen GN, Fang DD (2015) RNA-seq analysis of short fiber mutants Ligon-lintless-1 (Li 1) and - 2 (Li 2) revealed important role of aquaporins in cotton (Gossypium hirsutum L.) fiber elongation. BMC Plant Biol 15:65
Newman RH, Hill SJ, Harris PJ (2013) Wide-angle x-ray scattering and solid-state nuclear magnetic resonance data combined to test models for cellulose microfibrils in mung bean cell walls. Plant Physiol 163:1558–1567
Nie X, Huang C, You C, Li W, Zhao W, Shen C, Zhang B, Wang H et al (2016) Genome-wide SSR-based association mapping for fiber quality in nation-wide upland cotton inbreed cultivars in China. BMC Genomics 17:352
Paterson AH, Saranga Y, Menz M, Jiang CX, Wright RJ (2003) QTL analysis of genotype x environment interactions affecting cotton fiber quality. Theor Appl Genet 106:384–396

Pear JR, Kawagoe Y, Schreckengost WE, Delmer DP, Stalker DM (1996) Higher plants contain homologs of the bacterial celA genes encoding the catalytic subunit of cellulose synthase. Proc Natl Acad Sci U S A 93:12637–12642
Peret B, Li G, Zhao J, Band LR, Voss U, Postaire O, Luu DT, Da Ines O et al (2012) Auxin regulates aquaporin function to facilitate lateral root emergence. Nat Cell Biol 14:991–998
Postaire O, Tournaire-Roux C, Grondin A, Boursiac Y, Morillon R, Schaffner AR, Maurel C (2010) A PIP1 aquaporin contributes to hydrostatic pressure-induced water transport in both the root and rosette of Arabidopsis. Plant Physiol 152:1418–1430
Pu L, Li Q, Fan X, Yang W, Xue Y (2008) The R2R3 MYB transcription factor GhMYB109 is required for cotton fiber development. Genetics 180:811–820
Qin H, Guo W, Zhang YM, Zhang T (2008) QTL mapping of yield and fiber traits based on a four-way cross population in Gossypium hirsutum L. Theor Appl Genet 117:883–894
Qin YM, Hu CY, Pang Y, Kastaniotis AJ, Hiltunen JK, Zhu YX (2007) Saturated very-long-chain fatty acids promote cotton fiber and Arabidopsis cell elongation by activating ethylene biosynthesis. Plant Cell 19:3692–3704
Qin YM, Zhu YX (2011) How cotton fibers elongate: a tale of linear cell-growth mode. Cur Opin Plant Biol 14:106–111
Reinisch AJ, Dong J, Brubaker CL, Stelly DM, Wendel JF, Paterson AH (1994) A detailed RFLP map of cotton, Gossypium hirsutum X Gossypium barbadense - chromosome organization and evolution in a disomic polyploid genome. Genetics 138:829–847
Rong J, Pierce GJ, Waghmare VN, Rogers CJ, Desai A, Chee PW, May OL, Gannaway JR et al (2005) Genetic mapping and comparative analysis of seven mutants related to seed fiber development in cotton. Theor Appl Genet 111:1137–1146
Ruan YL (2005) Recent advances in understanding cotton fibre and seed development. Seed Sci Res 15:269–280
Ruan YL (2007) Rapid cell expansion and cellulose synthesis regulated by plasmodesmata and sugar: insights from the single-celled cotton fibre. Funct Plant Biol 34:1–10
Ruan YL, Chourey PS (1998) A fiberless seed mutation in cotton is associated with lack of fiber cell initiation in ovule epidermis and alterations in sucrose synthase expression and carbon partitioning in developing seeds. Plant Physiol 118:399–406
Ruan YL, Llewellyn DJ, Furbank RT (2001) The control of single-celled cotton fiber elongation by developmentally reversible gating of plasmodesmata and coordinated expression of sucrose and K+ transporters and expansin. Plant Cell 13:47–60
Ruan YL, Llewellyn DJ, Furbank RT, Chourey PS (2005) The delayed initiation and slow elongation of fuzz-like short fibre cells in relation to altered patterns of sucrose synthase expression and plasmodesmata gating in a lintless mutant of cotton. J Exp Bot 56:977–984
Rungis D, Llewellyn D, Dennis ES, Lyon BR (2005) Simple sequence repeat (SSR) markers reveal low levels of polymorphism between cotton (Gossypium hirsutum L.) cultivars. Aust J Agric Res 56:301–307
Said JI, Lin Z, Zhang X, Song M, Zhang J (2013) A comprehensive meta QTL analysis for fiber quality, yield, yield related and morphological traits, drought tolerance, and disease resistance in tetraploid cotton. BMC Genomics 14:776
Salih H, Gong W, He S, Sun G, Sun J, Du X (2016) Genome-wide characterization and expression analysis of MYB transcription factors in Gossypium hirsutum. BMC Genet 17:129
Seagull RW (1990) The effects of microtubule and microfilament disrupting agents on cytoskeletal arrays and wall deposition in developing cotton fbers. Protoplasma 159:44–59
Shan CM, Shangguan XX, Zhao B, Zhang XF, Chao LM, Yang CQ, Wang LJ, Zhu HY et al (2014) Control of cotton fibre elongation by a homeodomain transcription factor GhHOX3. Nat Commun 5:5519
Shangguan XX, Xu B, Yu ZX, Wang LJ, Chen XY (2008) Promoter of a cotton fibre MYB gene functional in trichomes of Arabidopsis and glandular trichomes of tobacco. J Exp Bot 59:3533–3542

Shangguan XX, Yang CQ, Zhang XF, Wang LJ (2016) Functional characterization of a basic helix-loop-helix (bHLH) transcription factor GhDEL65 from cotton (Gossypium hirsutum). Physiol Plant 158:200–212

Shen X, Cao Z, Singh R, Lubbers EL, Xu P, Smith CW, Paterson AH, Chee PW (2011) Efficacy of qFL-chr1, a quantitative trait locus for fiber length in cotton (Gossypium spp.). Crop Sci 51:2005–2010

Shi YH, Zhu SW, Mao XZ, Feng JX, Qin YM, Zhang L, Cheng J, Wei LP et al (2006) Transcriptome profiling, molecular biological, and physiological studies reveal a major role for ethylene in cotton fiber cell elongation. Plant Cell 18:651–664

Stewart MJ (1975) Fiber initiation on the cotton ovule (Gossypium Hirsutum). Am J Bot 62:723–730

Stiff MR, Tuttle JR, Graham BP, Haigler CH (2016) Cotton fiber biotechnology: potential controls and transgenic improvement of elongation and cell wall thickening. In: Ramawat KG, Ahuja MR (eds) Fiber plants: biology, biotechnology and applications. Springer, Cham, pp 127–153

Sun Z, Wang X, Liu Z, Gu Q, Zhang Y, Li Z, Ke H, Yang J et al (2017) Genome-wide association study discovered genetic variation and candidate genes of fibre quality traits in Gossypium hirsutum L. Plant Biotechnol J 15:982–996

Suo J, Liang X, Pu L, Zhang Y, Xue Y (2003) Identification of GhMYB109 encoding a R2R3 MYB transcription factor that expressed specifically in fiber initials and elongating fibers of cotton (Gossypium hirsutum L.). Biochim Biophys Acta 1630:25–34

Szymanski DB, Marks MD (1998) GLABROUS1 overexpression and TRIPTYCHON alter the cell cycle and trichome cell fate in Arabidopsis. Plant Cell 10:2047–2062

Taliercio EW, Boykin D (2007) Analysis of gene expression in cotton fiber initials. BMC Plant Biol 7:22

Tang W, He Y, Tu L, Wang M, Li Y, Ruan YL, Zhang X (2014a) Down-regulating annexin gene GhAnn2 inhibits cotton fiber elongation and decreases Ca2+ influx at the cell apex. Plant Mol Biol 85:613–625

Tang W, Tu L, Yang X, Tan J, Deng F, Hao J, Guo K, Lindsey K et al (2014b) The calcium sensor GhCaM7 promotes cotton fiber elongation by modulating reactive oxygen species (ROS) production. New Phytol 202:509–520

Thyssen GN, Fang DD, Turley RB, Florane C, Li P, Naoumkina M (2015) Mapping-by-sequencing of Ligon-lintless-1 (Li 1) reveals a cluster of neighboring genes with correlated expression in developing fibers of upland cotton (Gossypium hirsutum L.). Theor Appl Genet 128:1703–1712

Thyssen GN, Fang DD, Turley RB, Florane CB, Li P, Mattison CP, Naoumkina M (2017) A Gly65Val substitution in an actin, GhACT_LI1, disrupts cell polarity and F-actin organization resulting in dwarf, lintless cotton plants. Plant J 90:111–121

Thyssen GN, Fang DD, Zeng L, Song X, Delhom CD, Condon TL, Li P, Kim HJ (2016) The immature fiber mutant phenotype of cotton (Gossypium hirsutum) is linked to a 22-bp frame-shift deletion in a mitochondria targeted pentatricopeptide repeat gene. G3 (Bethesda) 6:1627–1633

Turley RB, Kloth RH (2002) Identification of a third fuzzless seed locus in upland cotton (Gossypium hirsutum L.). J Hered 9:359–364

Turley RB, Kloth RH (2008) The inheritance model for the fiberless trait in upland cotton (Gossypium hirsutum L.) line SL1-7-1: variation on a theme. Euphytica 164:123–132

Tuttle JR, Nah G, Duke MV, Alexander DC, Guan X, Song Q, Chen ZJ, Scheffler BE et al (2015) Metabolomic and transcriptomic insights into how cotton fiber transitions to secondary wall synthesis, represses lignification, and prolongs elongation. BMC Genomics 16:477

Vandavasi VG, Putnam DK, Zhang Q, Petridis L, Heller WT, Nixon BT, Haigler CH, Kalluri U et al (2016) A structural study of CESA1 catalytic domain of Arabidopsis cellulose synthesis complex: evidence for CESA trimers. Plant Physiol 170:123–135

Walford SA, Wu Y, Llewellyn DJ, Dennis ES (2011) GhMYB25-like: a key factor in early cotton fibre development. Plant J 65:785–797

Walford SA, Wu Y, Llewellyn DJ, Dennis ES (2012) Epidermal cell differentiation in cotton mediated by the homeodomain leucine zipper gene, GhHD-1. Plant J 71:464–478

Wan Q, Guan X, Yang N, Wu H, Pan M, Liu B, Fang L, Yang S et al (2016) Small interfering RNAs from bidirectional transcripts of GhMML3_A12 regulate cotton fiber development. New Phytol 210:1298–1310

Wan Q, Zhang H, Ye W, Wu H, Zhang T (2014) Genome-wide transcriptome profiling revealed cotton fuzz fiber development having a similar molecular model as Arabidopsis trichome. PLoS One 9:e97313

Wang B, Guo W, Zhu X, Wu Y, Huang N, Zhang T (2006) QTL mapping of fiber quality in an elite hybrid derived-RIL population of upland cotton. Euphytica 152:367–378

Wang B, Guo W, Zhu X, Wu Y, Huang N, Zhang T (2007) QTL mapping of yield and yield components for elite hybrid derived-RILs in upland cotton. J Genet Genom 34:35–45

Wang H, Guo Y, Lv F, Zhu H, Wu S, Jiang Y, Li F, Zhou B et al (2010a) The essential role of GhPEL gene, encoding a pectate lyase, in cell wall loosening by depolymerization of the de-esterified pectin during fiber elongation in cotton. Plant Mol Biol 72:397–406

Wang H, Huang C, Guo H, Li X, Zhao W, Dai B, Yan Z, Lin Z (2015) QTL mapping for fiber and yield traits in upland cotton under multiple environments. PLoS One 10:e0130742

Wang HY, Wang J, Gao P, Jiao GL, Zhao PM, Li Y, Wang GL, Xia GX (2009) Down-regulation of GhADF1 gene expression affects cotton fibre properties. Plant Biotechnol J 7:13–23

Wang J, Wang HY, Zhao PM, Han LB, Jiao GL, Zheng YY, Huang SJ, Xia GX (2010b) Overexpression of a profilin (GhPFN2) promotes the progression of developmental phases in cotton fibers. Plant Cell Physiol 51:1276–1290

Wang L, Cook A, Patrick JW, Chen XY, Ruan YL (2014) Silencing the vacuolar invertase gene GhVIN1 blocks cotton fiber initiation from the ovule epidermis, probably by suppressing a cohort of regulatory genes via sugar signaling. Plant J 78:686–696

Wang L, Li XR, Lian H, Ni DA, He YK, Chen XY, Ruan YL (2010c) Evidence that high activity of vacuolar invertase is required for cotton fiber and Arabidopsis root elongation through osmotic dependent and independent pathways, respectively. Plant Physiol 154:744–756

Wang MY, Zhao PM, Cheng HQ, Han LB, Wu XM, Gao P, Wang HY, Yang CL et al (2013) The cotton transcription factor TCP14 functions in auxin-mediated epidermal cell differentiation and elongation. Plant Physiol 162:1669–1680

Wang P, Zhu Y, Song X, Cao Z, Ding Y, Liu B, Zhu X, Wang S et al (2012) Inheritance of long staple fiber quality traits of Gossypium barbadense in G. hirsutum background using CSILs. Theor Appl Genet 124:1415–1428

Wang S, Wang JW, Yu N, Li CH, Luo B, Gou JY, Wang LJ, Chen XY (2004) Control of plant trichome development by a cotton fiber MYB gene. Plant Cell 16:2323–2334

Wu H, Tian Y, Wan Q, Fang L, Guan X, Chen J, Hu Y, Ye W et al (2018) Genetics and evolution of MIXTA genes regulating cotton lint fiber development. New Phytol 217:883–895

Wu Y, Llewellyn DJ, White R, Ruggiero K, Al-Ghazi Y, Dennis ES (2007) Laser capture microdissection and cDNA microarrays used to generate gene expression profiles of the rapidly expanding fibre initial cells on the surface of cotton ovules. Planta 226:1475–1490

Wu Y, Machado AC, White RG, Llewellyn DJ, Dennis ES (2006) Expression profiling identifies genes expressed early during lint fibre initiation in cotton. Plant Cell Physiol 47:107–127

Xiao YH, Li DM, Yin MH, Li XB, Zhang M, Wang YJ, Dong J, Zhao J et al (2010) Gibberellin 20-oxidase promotes initiation and elongation of cotton fibers by regulating gibberellin synthesis. J Plant Physiol 167:829–837

Xu B, Gou JY, Li FG, Shangguan XX, Zhao B, Yang CQ, Wang LJ, Yuan S et al (2013a) A cotton BURP domain protein interacts with alpha-expansin and their co-expression promotes plant growth and fruit production. Mol Plant 6:945–958

Xu P, Gao J, Cao Z, Chee PW, Guo Q, Xu Z, Paterson AH, Zhang X et al (2017) Fine mapping and candidate gene analysis of qFL-chr1, a fiber length QTL in cotton. Theor Appl Genet 130:1309–1319

Xu WL, Zhang DJ, Wu YF, Qin LX, Huang GQ, Li J, Li L, Li XB (2013b) Cotton PRP5 gene encoding a proline-rich protein is involved in fiber development. Plant Mol Biol 82:353–365

Yadav VK, Yadav VK, Pant P, Singh SP, Maurya R, Sable A, Sawant SV (2017) GhMYB1 regulates SCW stage-specific expression of the GhGDSL promoter in the fibres of Gossypium hirsutum L. Plant Biotechnol J 15:1163–1174

Yan Q, Wang Y, Li Q, Zhang Z, Ding H, Zhang Y, Liu H, Luo M et al (2018) Upregulation of GhTT2-3A in cotton fibers during secondary wall thickening results in brown fibers with improved quality. Plant Biotechnol J. https://doi.org/10.1111/pbi.12910

Yang SS, Cheung F, Lee JJ, Ha M, Wei NE, Sze SH, Stelly DM, Thaxton P et al (2006) Accumulation of genome-specific transcripts, transcription factors and phytohormonal regulators during early stages of fiber cell development in allotetraploid cotton. Plant J 47:761–775

Yu J, Kohel RJ, Smith CW (2010) The construction of a tetraploid cotton genome wide comprehensive reference map. Genomics 95:230–240

Zeng L, Meredith WR Jr, Gutierrez OA, Boykin DL (2009) Identification of associations between SSR markers and fiber traits in an exotic germplasm derived from multiple crosses among Gossypium tetraploid species. Theor Appl Genet 119:93–103

Zhang D, Hrmova M, Wan CH, Wu C, Balzen J, Cai W, Wang J, Densmore LD et al (2004) Members of a new group of chitinase-like genes are expressed preferentially in cotton cells with secondary walls. Plant Mol Biol 54:353–372

Zhang DY, Zhang TZ, Sang ZQ, Guo WZ (2007) Comparative development of lint and fuzz using different cotton fiber-specific developmental mutants in Gossypium hirsutum. J Integ Plant Biol 49:1038–1046

Zhang F, Jin X, Wang L, Li S, Wu S, Cheng C, Zhang T, Guo W (2016) A cotton annexin affects fiber elongation and secondary cell wall biosynthesis associated with Ca2+ influx, ROS homeostasis, and actin filament reorganization. Plant Physiol 171:1750–1770

Zhang F, Li S, Yang S, Wang L, Guo W (2015a) Overexpression of a cotton annexin gene, GhAnn1, enhances drought and salt stress tolerance in transgenic cotton. Plant Mol Biol 87:47–67

Zhang J, Huang GQ, Zou D, Yan JQ, Li Y, Hu S, Li XB (2018) The cotton (Gossypium hirsutum) NAC transcription factor (FSN1) as a positive regulator participates in controlling secondary cell wall biosynthesis and modification of fibers. New Phytol 217:625–640

Zhang K, Zhang J, Ma J, Tang S, Liu D, Teng Z, Liu D, Zhang Z (2012) Genetic mapping and quantitative trait locus analysis of fiber quality traits using a three-parent composite population in upland cotton (Gossypium hirsutum L.). Mol Breed 29:335–348

Zhang M, Han LB, Wang WY, Wu SJ, Jiao GL, Su L, Xia GX, Wang HY (2017a) Overexpression of GhFIM2 propels cotton fiber development by enhancing actin bundle formation. J Integr Plant Biol 59:531–534

Zhang M, Zheng X, Song S, Zeng Q, Hou L, Li D, Zhao J, Wei Y et al (2011) Spatiotemporal manipulation of auxin biosynthesis in cotton ovule epidermal cells enhances fiber yield and quality. Nature Biotechnol 29:453–458

Zhang TZ, Hu Y, Jiang W, Fang L, Guan X, Chen J, Zhang J, Saski CA et al (2015b) Sequencing of allotetraploid cotton (Gossypium hirsutum L. acc. TM-1) provides a resource for fiber improvement. Nature Biotechnol 33:531–537

Zhang TZ, Pan JJ (1991) Genetic analysis of a fuzzless-lintless mutant in Gossypium hirsutum L. Jiangsu J Agr Sci 7:13–16

Zhang TZ, Qian N, Zhu X, Chen H, Wang S, Mei H, Zhang Y (2013) Variations and transmission of QTL alleles for yield and fiber qualities in upland cotton cultivars developed in China. PLoS One 8:e57220

Zhang Y, He P, Yang Z, Huang G, Wang L, Pang C, Xiao H, Zhao P et al (2017b) A genome-scale analysis of the PIN gene family reveals its functions in cotton fiber development. Front Plant Sci 8:461

Zhang Z, Ge Q, Liu A, Li J, Gong J, Shang H, Shi Y, Chen T et al (2017c) Construction of a high-density genetic map and its application to QTL identification for fiber strength in upland cotton. Crop Sci 57:774–788

Zhang Z, Ruan YL, Zhou N, Wang F, Guan X, Fang L, Shang X, Guo W et al (2017d) Suppressing a putative sterol carrier gene reduces plasmodesmal permeability and activates sucrose transporter genes during cotton fiber elongation. Plant Cell 29:2027–2046

Zhao B, Cao JF, Hu GJ, Chen ZW, Wang LY, Shangguan XX, Wang LJ, Mao YB et al (2018) Core cis-element variation confers subgenome-biased expression of a transcription factor that functions in cotton fiber elongation. New Phytol 218:1061–1075

Zhu QH, Yuan Y, Stiller W, Jia Y, Wang P, Pan Z, Du X, Llewellyn D et al (2018) Genetic dissection of the fuzzless seed trait in Gossypium barbadense. J Exp Bot 69:997–1009

Zhu YQ, Xu KX, Luo B, Wang JW, Chen XY (2003) An ATP-binding cassette transporter GhWBC1 from elongating cotton fibers. Plant Physiol 133:580–588

Chapter 9
Advances in Understanding of Cotton Fiber Cell Differentiation and Elongation

Marina Naoumkina

9.1 Introduction

Cotton is the major renewable source of fibers used worldwide in the textile industry. Among about 50 species of the *Gossypium* genus, only four, that included two tetraploids (*Gossypium hirsutum* L. and *Gossypium barbadense* L.) and two diploids (*Gossypium herbaceum* L. and *Gossypium arboreum* L.), produce commercially important spinnable fiber. Cotton fibers are highly elongated single-celled seed trichomes that emerge from the outer epidermis of the ovule. Approximately one in four epidermal cells differentiates into a single-cell trichome during flowering. The development of fiber cells is quasi-synchronized and first occurs at the chalazal end of a cotton ovule (Steward 1975). Cotton fiber development consists of four distinct but overlapping stages, including initiation, elongation, secondary cell wall synthesis, and maturation (Kim and Triplett 2001). Fiber initiation is observed before or on the day of anthesis (Steward 1975). In cultivated cotton species, seed trichomes differentiate into two distinct types, long lint fibers that are easily detached from the epidermis and fuzz fibers that are short fibers strongly adhered to the seed coat (Applequist et al. 2001). The lint fiber cells elongate for about 27–39 days post anthesis (dpa), with secondary cell wall biosynthesis occurring from 17 to 53 dpa depending on the cotton variety and growth conditions (Benedict et al. 1999).

The density of lint fiber initials on the ovule surface determines the lint content on cotton seeds and may affect the fiber yield overall (Zhang et al. 2011), while the duration of the elongation period determines the length of lint fibers. The extent of secondary wall thickening controls fiber fineness. Increasing the fiber yield is of particular interest for cotton growers, while improving fiber length, strength, and

---

M. Naoumkina (✉)
Cotton Fiber Bioscience Research Unit, USDA-ARS, Southern Regional Research Center, New Orleans, LA, USA
e-mail: marina.naoumkina@ars.usda.gov

D. D. Fang (ed.), *Cotton Fiber: Physics, Chemistry and Biology*,
https://doi.org/10.1007/978-3-030-00871-0_9

optimal fineness is critical for the woven textile industry, particularly with regard to spinning, weaving, and knitting where rapid technological changes are being implemented. The presence or absence of fuzz fiber is an important trait for ginning cotton. The separation of the lint fiber from seeds by the roller gin is easier with seeds that have no fuzz fiber, and the number of seed coat neps and short fiber content is reduced in seeds without fuzz fibers (Bechere et al. 2011). Classical cotton breeding has been successful for years in improving fiber quality characteristics; however, plant breeding methods could not keep up with the required speed of fiber quality improvement to satisfy the rapid developments in the textile fiber processing industry (Benedict et al. 1999). Therefore, to stay competitive, the cotton fiber industry requires the development of new and efficient tools in breeding and biotechnology to improve fiber quality, which is impossible without fundamental knowledge of cotton fiber development. The basic understanding of cotton fiber development is still in its infancy; however, in recent years new information about genes controlling cotton fiber development has become available. This chapter provides an update on the current understanding of cotton fiber differentiation and elongation.

9.2 Genetic Control of Fuzz and Lint Fiber

Fiber mutants are valuable tools to study cotton fiber development. Currently, a number of mutants altered in fuzz and lint fiber development are described in the literature (Table 9.1). The images of some mutants described in this chapter are shown in Fig. 9.1. Lint initials develop on the day or before anthesis, while fuzz initials develop from a second wave a few days later. Lint fibers grow to lengths of approximately 30 mm, whereas fuzz fibers, for reasons yet to be determined, terminate elongation at approximately 5 mm in length (Zhang et al. 2007). A more thorough understanding of the molecular mechanisms controlling fuzz and lint fiber development may provide an avenue for fiber quality improvement through molecular breeding and biotechnology.

The presence or absence of fuzz fiber is believed to be determined by two main loci, which were found in both *G. hirsutum* and *G. barbadense* (Benedict et al. 1999; Endrizzi et al. 1984; Kearney and Harrison 1927; Ware et al. 1947). The dominant "naked seed" mutant that has no fuzz, but with lint on cotton seeds, controls the phenotype with a single dominant locus, N_l (Kearney and Harrison 1927). The second locus associated with fuzzless phenotype is recessive, n_2 (Ware et al. 1947). The N_l and n_2 loci reside on homeologous chromosomes A12 and D12, respectively (Endrizzi and Ramsay 1980; Wan et al. 2016). The dominant N_l allele has a more consistent naked seed phenotype and also affects lint fiber development, with lint varying extensively in N_l seeds from normal to nearly fiberless (Wan et al. 2016). The recessive n_2 allele produces either a completely or partially naked seed phenotype, depending on genetic background. The dominant allele has also been reported for N_2 locus, which removes all fuzz except for a tuft at the apical or funicular end of the seed (Endrizzi and Ray 1991).

Table 9.1 Cotton mutants with altered fiber differentiation and elongation

Fiber mutant	Phenotype	Reported loci	Reference
N_1	Fuzzless	N_1(GhA12G1503)	Kearney and Harrison (1927), Walford et al. (2011), and Wan et al. (2016)
n_2	Fuzzless	n_2n_3 Chr_D12	Endrizzi and Ramsay (1980), Turley and Kloth (2002), and Ware et al. (1947)
$9023n^t_4$	Tufted	n^t_4	Bechere et al. (2012)
MD17	Fiberless	$N_1n_2n_3$	Turley (2002) and Turley and Kloth (2002)
L-70	Fiberless	$If_{t1}f_{t2}f_c$	Musaev and Abzalov (1972)
SL1-7-1	Fiberless	$N_1fl_1n^3$	Turley and Kloth (2008)
MCU5	Fiberless	2–4 loci	Nadarajan and Rangasamy (1988)
Xu142*fl*	Fiberless	li_3n_2 (chr_D12)	Ma et al. (2016) and Zhang and Pan 1991
Li_1	Short fiber	GhD04G0865	Griffee and Ligon (1929) and Thyssen et al. (2017)
Li_x	Short fiber	chr_A04	Cai et al. (2013)
Li_2	Short fiber	Chr_D13	Hinchliffe et al. (2011), Kohel et al. (2002), Narbuth and Kohel (1990), Rong et al. (2005), and Thyssen et al. (2017)
li_y	Short fiber	NA	Naoumkina et al. (2017a)

Fig. 9.1 Open cotton bolls of *G. hirsutum* wild type (DP5690) and some mutant lines described in this chapter. The images for mutant lines Xu142*fl*, MD17, SL1-7-1 (Kim et al. 2015), N_1 (Kim and Triplett 2004), and Li_1 (Gilbert et al. 2013) were kindly provided by Dr. Kim. The images for li_y (Naoumkina et al. 2017a), wild-type, and Li_2 lines were created by the author

However, there is evidence that more genes may be involved in determination of fuzz fiber development. Musaev and Abzalov (1972) studied the fuzz inheritance in *G. hirsutum* seeds and suggested that at least four nonallelic genes (F_{t1}, F_{t2}, F_c, and *I*) and their interactions control the level of fuzz on cotton seeds. F_{t1} and F_{t2} determine fuzz on the micropyle region of seeds, and F_c determines fuzz on the chalazal region and lateral sides of seeds, whereas *I* (an inhibitor gene) in homozygous and heterozygous states blocks the function of these genes resulting in a dominant fuzzless phenotype. Turley and Kloth have detected a third recessive n_3 locus in the n_2 fuzzless mutant line (Turley and Kloth 2002). A new fuzzless recessive seed locus n^t_4 has been developed through chemical mutagenesis (Bechere et al. 2012). The n^t_4 seeds are partially naked (tufted) with fuzz grown around the micropyle end. The n^t_4 locus is not allelic to N_1 or n_2. Unlike in N_1 and n_2, the lint percentage in the n^t_4 mutant was not significantly lower compared to the wild type.

Interactions between fuzzless loci to produce a fiberless phenotype are particularly interesting. A few fiberless (no fuzz and no lint) mutants have been identified throughout the world (Table 9.1). Among them are SL1-7-1 fiberless (USA, Fig. 9.1), MCU5 fiberless (India), L-70 fiberless (Uzbekistan), and Xu142*fl* (China, Fig. 9.1). The fiberless seed phenotype of SL1-7-1 is expressed by three loci, including N_1, fl_1, and n_3 (Turley and Kloth 2008). A genetic study of the MCU5 fiberless line suggested that its character is controlled by two to four gene pairs (Nadarajan and Rangasamy 1988); and speculations have been made based on the reported segregation of F_2 progeny that the MCU5 line has either the n_2 or n_3 locus along with other loci (Turley and Kloth 2008). In the L-70 fiberless line, the dominant gene (*I*) epistatically inhibits the expression of fuzz (Musaev and Abzalov 1972); it was speculated later that the inhibitor gene could be equivalent to the dominant N_2 locus, since naked seeds have a tufted phenotype (Du et al. 2001). The fiberless character of Xu142*fl* has been shown to be controlled by two loci, li_3 and n_2, both located on chromosome D12 (Ma et al. 2016; Zhang and Pan 1991). All fiberless mutants contain a fuzzless locus, either N_1 or n_2, along with other gene modifiers. Turley (2002) experimentally demonstrated the interactions of mutated loci, responsible for fuzzless phenotype N_1 and n_2 to produce fiberless seed. The fiberless line MD17 (Fig. 9.1) was developed by crossing two fuzzless mutants, N_1 and n_2 (Turley 2002). Epistatic interactions between fuzz and lint had been genetically analyzed by Du and coauthors (Du et al. 2001). Fuzzless and fiberless mutants were crossed with each other and wild type to produce F_2 and BC_1 generations. From all the expected fiber combinations, plants forming fuzzy-lintless seeds were not observed in experiment nor are known in nature.

From the review of these studies, we can observe that there are multiple loci involved in the fuzz fiber development. Interactions of fuzzless loci produce a fiberless seed phenotype; therefore, fuzz and lint fiber share common regulators, but different mechanisms of actions since fuzz fibers stop elongation earlier than lint fibers. The lint fiber development is epistatic to fuzz since there are no known fuzzy-lintless mutants.

9.3 Transcriptional Regulation of Fiber Development

9.3.1 Fuzz and Lint Differentiation

Recent progress has been made in understanding on molecular level fuzz and lint fiber differentiation processes. The mutation responsible for naked N_l seed fiber phenotype is in a MYBMIXTA-like transcription factor 3 [MML3, other name GhMYB25-like (Wan et al. 2016)], gene number GhA12G1503 in the *G. hirsutum* TM-1 draft genome (Zhang et al. 2015). The GhMML3 genomic sequence has a 1161 bp deletion fragment in the second intron. It is unclear whether this fragment loss or another factor triggers the natural antisense (NAT) production driven by its 3′ antisense promoter that significantly reduces expression of both homeologs of GhMML3 in N_l plants. Small RNA analysis revealed that bidirectional transcription of GhMML3 generated siRNAs. The siRNA-mediated GhMML3 mRNA self-cleavage could potentially play a role as a transfactor that regulates downstream genes. The GhMML and GhMYB transcription factors have complementary sequences that can be targeted by siRNAs derived from GhMML3 (Wan et al. 2016). A similar small RNA distribution was found in two fiberless mutants containing the N_l locus (SL1-7-1 and MD17) and in F_l plants derived from a cross of N_l × CSIL158, indicating stable inheritance of this regulation.

A previously published independent study demonstrated a critical role of the same gene GhMYB25-like (GhMML3) in early cotton fiber development (Walford et al. 2011). The authors of this study suppressed the expression of both homeologs of GhMML3 through RNA interference (RNAi) which resulted in fiberless seeds but normal trichomes elsewhere, reproducing phenotypes of the Xu142*fl* fiberless mutant (Walford et al. 2011). The natural N_l mutation and artificial RNAi technique both induced siRNA production that cleaved GhMML3 homeologs and probably other homologous genes.

The causative nature of another fuzzless mutation n_2 is probably different. Whether the n_2 gene is encoded by homeologous GhMML3 on chromosome D12 still requires confirmation, but everything suggests this conclusion. The n_2 mutated locus has been mapped to D12, which is the homeologous chromosome of N_l gene (Ma et al. 2016). A point mutation which caused a nonconservative amino acid substitution in the highly conserved R2R3 MYB domain was detected in GhMML3 of Xu142*fl* (Walford et al. 2011). The alterations in this region of consensus usually lead to reduced or abolished DNA-binding activity in other MYBs (Serna and Martin 2006). Such obliteration of the DNA-binding activity of the GhMML3 protein would produce a similar effect as the absence of GhMML3 protein due to RNAi-induced silencing; however, this effect would not be dominant. The presence of a homeologous copy might be enough to partially maintain the function and produce lint fibers which are observed in the n_2 mutant (Ware et al. 1947).

Genetic studies of fiberless mutants suggested the involvement of at least two genes in lint fiber development, always including the fuzzless loci (N_l or n_2) with gene modifiers (Du et al. 2001; Turley and Kloth 2002). For example, in the natural mutant Xu142*fl*, the n_2 and li_3 loci are responsible for the fiberless phenotype, while in SL1-7-1 three loci, N_1, fl_1, and n_3, are involved in fiberless seed production. We know now that N_1 is encoded by MML3 and suggest n_2 is encoded by the MML3 homeolog on chromosome D12. It was recently discovered that the gene responsible for li_3 locus in Xu142*fl* mutant is also encoded by MIXTA-type MYB transcription factor GhMML4 located on chromosome D12 (Wu et al. 2017). The GhMML4 homeolog is not transcriptionally active during fiber initiation, suggesting specialization of MML4 homeologs (Wu et al. 2017). The GhMML4 gene is positioned next to GhMML3 on chromosome D12 in the *G. hirsutum* genome and is critical for lint fiber production; however, to produce the fiberless phenotype, both genes should be disabled.

The puzzle is how the fiberless phenotype works in MD17 (Turley 2002). Both N_1 and n_2 mutants have no fuzz but produce lint fiber, while crossing them resulted in a fiberless phenotype. The partial answer may lie in differences of transcriptional activity between GhMML3 homeologs. The GhMML3 A12 homeolog reached maximum transcript level at 1 DPA, while the GhMML3 D12 homeolog transcription peaked earlier at −1 DPA (Wu et al. 2017). In the N_1 mutant, the highest siRNA activity correlated with transcriptional activity of the GhMML3 A12 gene at 1 DPA, the time point it cleaved transcripts of both homeologs at maximal rate. However, functional activity of the other homeolog GhMML3 D12 from −1 to 1 DPA might be enough to initiate lint fiber development (fuzz fiber differentiates later); therefore, N_1 mutants produce lint fiber, still with great a degree of variation in lint percentage. In n_2 mutants, the GhMML3 A12 maintains function for lint production, and we see variations in the amount of fuzz on n_2 seeds. The combination of two impaired genes (N_1 and n_2) produces a fiberless phenotype. The same effect is seen by constitutive RNAi suppression of GhMML3 genes (Walford et al. 2011). There are still more questions than answers. For example, Xu142*fl* mutant has an intact copy of GhMML3 A12 but still produces a fiberless phenotype. This means that GhMML4 D12 is the next critical link in the fiber developmental process. The GhMML4 D12 li_3 allele has a premature stop codon in the C-terminal domain of the protein, but an intact R2R3 MYB binding domain (Wu et al. 2017). If a truncated GhMML4 D12 protein is produced, it would likely bind to available GhMML3 proteins and block the subsequent transcriptional cascade for fuzz and lint fiber development. However, this suggestion still requires experimental evaluation.

9.3.2 Transcriptional Network

Now, a preliminary model for fiber development is emerging with a critical role in this process for MIXTA MYB transcription factors (Fig. 9.2). The GhMML3 homeologs are the initial point in fuzz and lint fiber development (Bedon et al. 2014; Walford et al. 2011; Wan et al. 2016; Wu et al. 2017). Interaction of GhMML3 with GhMML4 is required for lint fiber development (Wu et al. 2017). However,

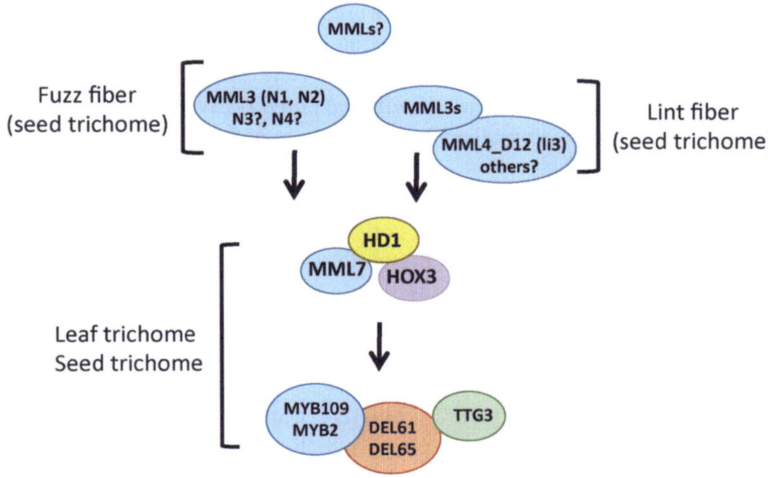

Fig. 9.2 A preliminary model of a transcription factor network for early fiber development. The top part of the figure depicts proposed transcription factors involved in seed trichome development (fuzz or lint fiber), while the next part shows proposed transcription factors common for both seed and leaf trichome development. Detailed explanation provided in the transcriptional network section of the main text

existence of other fuzzless and lintless loci (n_3, n_4, fl_1, etc.) suggests that fiber initiation has a more complex transcriptional network.

There are ten GhMML family members detected in the *G. hirsutum* genome (Zhang et al. 2015). Almost all of them are expressed in early developing ovules, and a few (GhMML5, 6, and 10) are expressed in leaves and/or cotyledons which are rich in trichomes (Bedon et al. 2014). GhMML7 (GhMYB25) was identified as a differentially expressed gene between a fiberless mutant and wild-type cotton plants (Lee et al. 2006; Wu et al. 2006). Ectopic expression of GhMML7 in cotton increased fiber initials and leaf trichome number, while RNAi silencing reduced fiber length and trichome number elsewhere (Machado et al. 2009), suggesting that GhMML7 is involved in seed and leaf trichome development. On the other hand, silencing of GhMML3 in transgenic cotton resulted in fiberless seeds, but normal trichomes elsewhere, whereas overexpressing an additional copy of GhMML3 did not increase fiber initials, nor did it affect trichomes on other parts of plant. This indicates a specific involvement of GhMML3 in seed trichome development (Walford et al. 2011).

To test the interaction network of fiber MYBs, the expression levels of different GhMMLs were evaluated in outer-integuments of developing ovules from −4 to 2 DPA collected from GhMML3 or GhMML7 silenced transgenic cotton plants (Bedon et al. 2014; Machado et al. 2009; Walford et al. 2011). In the GhMML3 silenced plants, the transcript levels of GhMML3, 7, 8, 9, and 4 were abolished or significantly reduced, while levels of GhMML2 and 6 were significantly higher. In the GhMML7 silenced plants, the transcript level of GhMML7 was reduced, while transcript levels of GhMML3, 4, 8, and 9 were elevated, indicating some feedback regulation of GhMML3 by GhMML7 (Bedon et al. 2014). Transactivation of the GhMML7 promoter, fused to the luciferase reporter, did occur in the presence of GhMML3, but only when co-expressed with the GhHD1 (Walford et al. 2012), suggesting that the two may be involved in a complex to activate GhMML7.

HD1 is a homeodomain-leucine zipper transcription factor and plays important roles in seed and stem trichome development. In *G. barbadense* a retrotransposon insertion into the GbHD1 A subgenome copy resulted in the loss of stem trichomes (Ding et al. 2015). Silencing of both homeologs of GhHD1 by RNAi reduced trichome formation and delayed fiber initiation, while overexpression increased fiber initials but did not affect trichomes elsewhere (Walford et al. 2012). GhHD1 expression was reduced in GhMML3 RNAi silenced plants, as well as in the Xu142*fl* fiberless mutant (where GhMML3 expression was also low) suggesting this gene is downstream of GhMML3 (Fig. 9.2). GhHD1 transcription was unchanged in GhMML7 and GhMYB109 silenced plants which produced almost normal fibers (Walford et al. 2012). However, expression levels of GhMML7 and HOX3 were significantly reduced in the retrotransposon-interrupted GhHD1 plants indicating that GhHD1 is upstream of GhMML7 and HOX3 (Ding et al. 2015). Taken together, these results suggest that GhHD1 is an important molecular switch that can activate GhMML7 and HOX3 during fiber and other trichome development (Bedon et al. 2014; Ding et al. 2015).

HOX3 is a homeodomain-leucine zipper IV (HD-ZIP IV) transcription factor involved in regulation of fiber elongation. Silencing GhHOX3 reduced fiber length up to 80% and reduced trichomes on stems and leaf veins, while overexpression resulted in a fiber length increase of up to 20% (Shan et al. 2014). GhHD1 is among the proteins that interact with GhHOX3 that were identified by a yeast two-hybrid assay. The expression of GhHD1 was not affected in GhHOX3 suppression lines suggesting that GhHD1 is upstream of GhHOX3 (Fig. 9.2).

In Arabidopsis, the Gl1-Gl3/Egl3-TTG1 complex plays a central regulatory role in trichome formation (Ishida et al. 2008). In cotton fiber development, this complex also exists, but functions downstream of MIXTA MYB transcription factors. For example, GhMYB109, the closest homolog of AtGl1, or AtWER, is not required for fiber initiation but important for fiber elongation. Antisense transgenic plants of GhMYB109 showed about 8% reduction in fiber initials and 33% reduction in fiber length (Pu et al. 2008). GhDEL65/GhDEL61, bHLH family proteins in cotton, are functional homologs of AtGl3 or AtEGl3. Downregulation of GhDEL65 resulted in slightly shorter fibers, while overexpression slightly increased fiber length (Shangguan et al. 2016). No significant effect on leaf or stem trichomes was observed in transgenic lines. GhDEL65 interacts with GhMYB2 or GhMYB3 (homologs of GhMYB109) as well as GhTTG3 in yeast cells. A preliminary model for cotton fiber development is proposed in Fig. 9.2. The model shows only transcription factors whose involvement was confirmed by mutation or demonstrated in transgenic cotton. It is clear that GhMML3 is a key factor in initial cotton fiber development, whereas the downstream chain of interactions among transcription factors is still obscure and will require extensive studies.

9.4 Fiber Elongation Mutants

There are several fiber elongation-related mutants described in the literature, including the dominant Li_1, Li_x, and Li_2 and recessive li_y (Table 9.1). All these mutants exhibit a short fiber phenotype, and three of them, Li_1, Li_x, and li_y, also show a pleiotropic effect on vegetative growth. The vegetative phenotype of Li_1 and Li_x is characterized by dwarf plants with twisted stems and crinkled leaves, whereas homozygous li_y plants are shorter, reaching approximately 50% of wild-type height.

The causative mutation of Li_1 (Fig. 9.1) is a substitution of Gly to Val at position 65 in the protein sequence of an actin gene Gh_D04G0865 (Thyssen et al. 2017). This actin isoform point mutation likely disrupts processive elongation of F-actin, resulting in a disorganized cytoskeleton and reduced cell polarity and consequently dwarf, lintless cotton plants (Thyssen et al. 2017). The importance of the actin cytoskeleton for cotton fiber elongation has been shown before. The cotton *ACTIN1* (AY305723, Gh_A11G0512) predominantly expressed in fiber cells was suppressed by RNAi that resulted in inhibition of early fiber elongation (Li et al. 2005).

The Li_x mutation occurred during a tissue culture process and is not related to a T-DNA insertion (Cai et al. 2013). The Li_x vegetative phenotype resembles Li_1; however, homozygous Li_x plants do not produce seeds, while fiber on heterozygous Li_x seeds is much longer than on heterozygous Li_1 seeds (Cai et al. 2013). The Li_x locus was mapped on chromosome A04 (homeologous to D04 of Li_1), though allelic analysis of Li_x/Li_1 crosses determined that Li_x and Li_1 were nonallelic (Cai et al. 2013). It would be interesting to see which gene was altered and the nature of the causative mutation of Li_x. Recently, a chemically induced short fiber mutant li_y, which is controlled by a single recessive locus, was reported (Naoumkina et al. 2017a). The mutation affects multiple traits, including height of the plant and length and maturity of fiber (Naoumkina et al. 2017a). The causative mutation of li_y has yet to be discovered.

The most interesting mutation is Li_2 (Fig. 9.1) since it does not cause adverse effects to plant growth and development and only visibly alters the lint fibers. The Li_2 locus was mapped on chromosome D13 (Hinchliffe et al. 2011; Kohel et al. 2002; Rong et al. 2005; Thyssen et al. 2014); however, the causative gene is still unknown. Extensive comparative studies of developing fibers of Li_2/wild-type near-isogenic lines detected global metabolic and transcriptional changes in response to the Li_2 mutation (Hinchliffe et al. 2011; Naoumkina et al. 2013). The effect of the Li_2 mutation was explored on homeolog expression bias in allotetraploid cotton. The results showed that the significant reduction of homeolog expression bias in mutant fiber correlates with negative fiber traits, indicating that the extent of homeolog expression bias is important for fiber quality characteristics (Naoumkina et al. 2014). Studies were performed to determine similarities in mechanisms of cessation of fiber elongation between Li_1 and Li_2. Small RNA sequencing and degradome analysis of developing fibers of short fiber mutants Li_1 and Li_2 revealed a role for miRNAs and their targets in cotton fiber elongation (Naoumkina et al. 2016). Elevated accumulation of xyloglucan was detected in Li_1 and Li_2 developing fibers that may restrict elongation of fiber cells (Naoumkina et al. 2017b). The major intrinsic proteins were found among the most downregulated gene family in both *Li₁* and *Li₂* fiber mutants (Naoumkina et al. 2015). The osmolality and concentrations of soluble sugars were lower in saps of *Li₁–Li₂* fibers, whereas the concentrations of malic acid, K+, and other detected ions were significantly higher in saps of fibers from mutants than in WT (Naoumkina et al. 2015). These results suggest that a higher accumulation of ions in fiber cells, reduced osmotic pressure, and low expression of aquaporins may contribute to the cessation of fiber elongation in *Li₁* and *Li₂* short fiber mutants (Naoumkina et al. 2015).

9.5 Future Perspectives

In the past decades, substantial progress has been made in detecting the genes that control fiber development in cotton. However, much remains unknown and needs to be elucidated in future research. Analyses of fiber mutants have been instrumental in better understanding the cotton fiber developmental processes. The causative mutations were discovered in only a few cotton fiber mutant lines. There are many more that remain to be revealed, and each will provide an essential contribution toward understanding the mechanism of cotton fiber development. We know now that GhMML transcription factors play an important role in cotton fiber development. Among ten GhMMLs in the *G. hirsutum* genome, only a few of them have been characterized; functional analyses are required for the rest. The functional characterization of potential fiber regulatory genes by silencing and overexpression in transgenic cotton are reasonable approaches. However, RNAi or antisense silencing sometimes may unintentionally target both homeologs and other members of a gene family. Therefore, another technique such as CRISPR/Cas9 should be considered in knockdown (or alteration) of specific homeologous genes. This is important especially if neofunctionalization of homeologous genes is suspected. The studies of protein-protein interactions in both yeast and cotton protoplasts will be helpful to understand the regulatory network.

Acknowledgment I greatly appreciate Drs. Gregory Thyssen and Rick Turley and two anonymous reviewers' constructive comments and suggestions, which helped to improve this review. I am also grateful to Dr. Hee Jin Kim for creating and sharing the images of cotton fiber mutants.

References

Applequist WL, Cronn R, Wendel JF (2001) Comparative development of fiber in wild and cultivated cotton. Evol Dev 3:3–17

Bechere E, Boykin JC, Meredith W (2011) Evaluation of cotton genotypes for ginning energy and ginning rate. J Cotton Sci 15:11–21

Bechere E, Turley RB, Auld DL, Zeng L (2012) A new fuzzless seed locus in an upland cotton (*Gossypium hirsutum* L.) mutant. Am J Plant Sci 3:799

Bedon F, Ziolkowski L, Walford SA, Dennis ES, Llewellyn DJ (2014) Members of the MYBMIXTA-like transcription factors may orchestrate the initiation of fiber development in cotton seeds. Front Plant Sci 5:179

Benedict CR, Kohel RJ, Lewis HL (1999) Cotton fiber quality. In: Smith CW, Cothren JT (eds) Cotton: origin, history, technology, and production. Wiley, New York, pp 269–288

Cai C, Tong X, Liu F, Lv F, Wang H, Zhang T, Guo W (2013) Discovery and identification of a novel Ligon lintless-like mutant (Li_x) similar to the Ligon lintless (Li_I) in allotetraploid cotton. Theor Appl Genet 126:963–970

Ding M, Ye W, Lin L, He S, Du X, Chen A, Cao Y, Qin Y et al (2015) The hairless stem phenotype of cotton (*Gossypium barbadense*) is linked to a copia-like retrotransposon insertion in a homeodomain-leucine zipper gene (HD1). Genetics 201:143–154

Du X, Pan J, Wang R, Zhang TZ, Shi YZ (2001) Genetic analysis of presence and absence of lint and fuzz in cotton. Plant Breed 120:519–522

Endrizzi J, Ray D (1991) Monosomic and monotelodisomic analysis of 34 mutant loci in cotton. J Hered 82:53–57

Endrizzi J, Turcotte E, Kohel R (1984) Qualitative genetics, cytology, and cytogenetics. In: Kohel R, Lewis F (eds) Cotton. ASA-CSSA-SSSA, Madison, WI, pp 81–129

Endrizzi JE, Ramsay G (1980) Identification of ten chromosome deficiencies of cotton. Cytological identification of eight chromosomes and genetic analysis of chromosome deficiencies and marker genes. J Hered 71:45–48

Gilbert MK, Turley RB, Kim HJ, Li P, Thyssen G, Tang Y, Delhom CD, Naoumkina M et al (2013) Transcript profiling by microarray and marker analysis of the short cotton (*Gossypium hirsutum* L.) fiber mutant Ligon lintless-1 (Li1). BMC Genomics 14:403

Griffee F, Ligon L (1929) Occurrence of "lintless" cotton plants and the inheritance of the character "lintless". J Am Soc Agron 21:711–717

Hinchliffe DJ, Turley RB, Naoumkina M, Kim HJ, Tang Y, Yeater KM, Li P, Fang DD (2011) A combined functional and structural genomics approach identified an EST-SSR marker with complete linkage to the Ligon lintless-2 genetic locus in cotton (*Gossypium hirsutum* L.). BMC Genomics 12:445

Ishida T, Kurata T, Okada K, Wada T (2008) A genetic regulatory network in the development of trichomes and root hairs. Annu Rev Plant Biol 59:365–386

Kearney TH, Harrison GJ (1927) Inheritance of smooth seeds in cotton. J Agric Res 35:193–217

Kim HJ, Hinchliffe DJ, Triplett BA, Chen ZJ, Stelly DM, Yeater KM, Moon HS, Gilbert MK et al (2015) Phytohormonal networks promote differentiation of fiber initials on pre-anthesis cotton ovules grown in vitro and in planta. PLoS One 10:e0125046

Kim HJ, Triplett BA (2001) Cotton fiber growth in planta and in vitro. Models for plant cell elongation and cell wall biogenesis. Plant Physiol 127:1361–1366

Kim HJ, Triplett BA (2004) Cotton fiber germin-like protein. I. Molecular cloning and gene expression. Planta 218:516–524

Kohel R, Stelly D, Yu J (2002) Tests of six cotton (*Gossypium hirsutum* L.) mutants for association with aneuploids. J Hered 93:130–132

Lee JJ, Hassan OS, Gao W, Wei NE, Kohel RJ, Chen X-Y, Payton P, Sze S-H et al (2006) Developmental and gene expression analyses of a cotton naked seed mutant. Planta 223:418–432

Li XB, Fan XP, Wang XL, Cai L, Yang WC (2005) The cotton ACTIN1 gene is functionally expressed in fibers and participates in fiber elongation. Plant Cell 17:859–875

Ma Q-F, Wu C-H, Wu M, Pei W-F, Li X-L, Wang W-K, Zhang J, Yu J-W et al (2016) Integrative transcriptome, proteome, phosphoproteome and genetic mapping reveals new aspects in a fiberless mutant of cotton. Sci Rep 6:24485

Machado A, Wu Y, Yang Y, Llewellyn DJ, Dennis ES (2009) The MYB transcription factor GhMYB25 regulates early fibre and trichome development. Plant J 59:52–62

Musaev D, Abzalov M (1972) Some questions concerning the inheritance of fuzzy in cotton seeds (*G. hirsutum* L.). Genetika 8:7–16

Nadarajan N, Rangasamy SS (1988) Inheritance of the fuzzless-lintless character in cotton (*Gossypium hirsutum*). Theor Appl Genet 75:728–730

Naoumkina M, Bechere E, Fang DD, Thyssen GN, Florane CB (2017a) Genome-wide analysis of gene expression of EMS-induced short fiber mutant Ligon lintless-y (liy) in cotton (*Gossypium hirsutum* L.). Genomics 109:320–329

Naoumkina M, Hinchliffe DJ, Fang DD, Florane CB, Thyssen GN (2017b) Role of xyloglucan in cotton (Gossypium hirsutum L.) fiber elongation of the short fiber mutant Ligon lintless-2 (Li2). Gene 626:227–233

Naoumkina M, Hinchliffe DJ, Turley RB, Bland JM, Fang DD (2013) Integrated metabolomics and genomics analysis provides new insights into the fiber elongation process in Ligon lintless-2 mutant cotton (*Gossypium hirsutum* L.). BMC Genomics 14:155

Naoumkina M, Thyssen G, Fang DD, Hinchliffe DJ, Florane C, Yeater KM, Page JT, Udall JA (2014) The Li_2 mutation results in reduced subgenome expression bias in elongating fibers of allotetraploid cotton (*Gossypium hirsutum* L.). PLoS One 9:e90830

Naoumkina M, Thyssen GN, Fang DD (2015) RNA-seq analysis of short fiber mutants Ligon-lintless-1 (Li_1) and −2 (Li_2) revealed important role of aquaporins in cotton (*Gossypium hirsutum* L.) fiber elongation. BMC Plant Biol 15:65

Naoumkina M, Thyssen GN, Fang DD, Hinchliffe DJ, Florane CB, Jenkins JN (2016) Small RNA sequencing and degradome analysis of developing fibers of short fiber mutants Ligon-lintles-1 (Li_1) and −2 (Li_2) revealed a role for miRNAs and their targets in cotton fiber elongation. BMC Genomics 17:360

Narbuth E, Kohel R (1990) Inheritance and linkage analysis of a new fiber mutant in cotton. J Hered 81:131–133

Pu L, Li Q, Fan X, Yang W, Xue Y (2008) The R2R3 MYB transcription factor GhMYB109 is required for cotton fiber development. Genetics 180:811–820

Rong J, Pierce GJ, Waghmare VN, Rogers CJ, Desai A, Chee PW, May OL, Gannaway JR et al (2005) Genetic mapping and comparative analysis of seven mutants related to seed fiber development in cotton. Theor Appl Genet 111:1137–1146

Serna L, Martin C (2006) Trichomes: different regulatory networks lead to convergent structures. Trends Plant Sci 11:274–280

Shan C-M, Shangguan X-X, Zhao B, Zhang X-F, L-m C, Yang C-Q, Wang L-J, Zhu H-Y et al (2014) Control of cotton fibre elongation by a homeodomain transcription factor GhHOX3. Nat Commun 5:5519

Shangguan XX, Yang CQ, Zhang XF, Wang LJ (2016) Functional characterization of a basic helix-loop-helix (bHLH) transcription factor GhDEL65 from cotton (*Gossypium hirsutum*). Physiol Plant 158:200–212

Steward JM (1975) Fiber initiation on the cotton ovule (*Gossypium hirsutum*). Am J Bot 62:723–730

Thyssen GN, Fang DD, Turley RB, Florane C, Li P, Naoumkina M (2014) Next generation genetic mapping of the Ligon-lintless-2 (Li_2) locus in upland cotton (*Gossypium hirsutum* L.). Theor Appl Genet 127:2183–2192

Thyssen GN, Fang DD, Turley RB, Florane CB, Li P, Mattison CP, Naoumkina M (2017) A Gly65Val substitution in an actin, GhACT_LI1, disrupts cell polarity and F-actin organization resulting in dwarf, lintless cotton plants. Plant J 90:111–121

Turley R (2002) Registration of MD 17 fiberless upland cotton as a genetic stock. (Registrations of genetic stocks). Crop Sci 42:994–996

Turley R, Kloth R (2002) Identification of a third fuzzless seed locus in upland cotton (*Gossypium hirsutum* L.). J Hered 93:359–364

Turley RB, Kloth RH (2008) The inheritance model for the fiberless trait in upland cotton (*Gossypium hirsutum* L.) line SL1-7-1: variation on a theme. Euphytica 164:123–132

Walford SA, Wu Y, Llewellyn DJ, Dennis ES (2011) GhMYB25-like: a key factor in early cotton fibre development. Plant J 65:785–797

Walford SA, Wu Y, Llewellyn DJ, Dennis ES (2012) Epidermal cell differentiation in cotton mediated by the homeodomain leucine zipper gene, GhHD-1. Plant J 71:464–478

Wan Q, Guan X, Yang N, Wu H, Pan M, Liu B, Fang L, Yang S et al (2016) Small interfering RNAs from bidirectional transcripts of GhMML3_A12 regulate cotton fiber development. New Phytol 210:1298–1310

Ware JO, Benedict LI, Rolfe WH (1947) A recessive naked-seed character in upland cotton. J Hered 38:313–320

Wu H, Tian Y, Wan Q, Fang L, Guan X, Chen J, Hu Y, Ye W et al (2017) Genetics and evolution of MIXTA genes regulating cotton lint fiber development. New Phytol 217:883–895

Wu Y, Machado AC, White RG, Llewellyn DJ, Dennis ES (2006) Expression profiling identifies genes expressed early during lint fibre initiation in cotton. Plant Cell Physiol 47:107–127

Zhang D, Zhang T, Sang Z, Guo W (2007) Comparative development of lint and fuzz using different cotton fiber-specific developmental mutants in *Gossypium hirsutum*. J Integ Plant Biol 49:1038–1046

Zhang M, Zheng X, Song S, Zeng Q, Hou L, Li D, Zhao J, Wei Y et al (2011) Spatiotemporal manipulation of auxin biosynthesis in cotton ovule epidermal cells enhances fiber yield and quality. Nat Biotechnol 29:453–458

Zhang T, Hu Y, Jiang W, Fang L, Guan X, Chen J, Zhang J, Saski CA et al (2015) Sequencing of allotetraploid cotton (*Gossypium hirsutum* L. acc. TM-1) provides a resource for fiber improvement. Nat Biotechnol 33:531–537

Zhang TZ, Pan JJ (1991) Genetic analysis of fuzzless-lintless mutant in upland cotton. Jiangsu J Agric Sci 7:14–16

Chapter 10
Cotton Fiber Improvement Through Breeding and Biotechnology

B. Todd Campbell, Jane K. Dever, Kari L. Hugie, and Carol M. Kelly

10.1 Introduction

Genetic improvement of upland cotton (*Gossypium hirsutum* L.) fiber quality is essential to meet demands of textile manufacturers and the international market while also increasing competitiveness with synthetic fibers. Changes in spinning and weaving technologies over the past two decades have increased the need for upland cotton with improved fiber qualities, particularly longer and more uniform fibers (Bourland and Jones 2012; Cantrell et al. 2000; Meredith and Nokes 2011; Smith et al. 2008, 2009). In comparison to synthetic fibers, demand for cotton fiber over the last 5 years has changed from 50% cotton and 44% synthetic in 2011 to 43% cotton and 50% synthetic in 2016 (Meyer 2017).

Cotton merchants and textile manufacturers are concerned with product quality. The specific needs of textile mills are defined by the end product, the yarn type, and the equipment or technology available. However, any and all textile manufacturing can be demanding on the fiber, and processes such as spinning, weaving, and dyeing are all affected by various fiber properties (Calhoun and Bowman 1999). Advances in technology have led to an increase in processing speeds thus placing more strain on cotton fibers (Benedict et al. 1999; May and Jividen 1999). This trend will most likely continue and result in a continuing need for improved cotton fibers that can perform under the new standards and withstand higher processing speeds.

Certain spinning techniques may be more affected by overall fiber quality or even be sensitive to specific fiber properties. For example, ring spinning generally requires a higher-quality fiber than is needed for rotor spinning. This is because of

B. T. Campbell (✉) · K. L. Hugie
USDA-ARS Coastal Plains Soil, Water, and Plant Research Center, Florence, SC, USA
e-mail: todd.campbell@ars.usda.gov

J. K. Dever · C. M. Kelly
Texas A&M AgriLife Research, Lubbock, TX, USA

D. D. Fang (ed.), *Cotton Fiber: Physics, Chemistry and Biology*,
https://doi.org/10.1007/978-3-030-00871-0_10

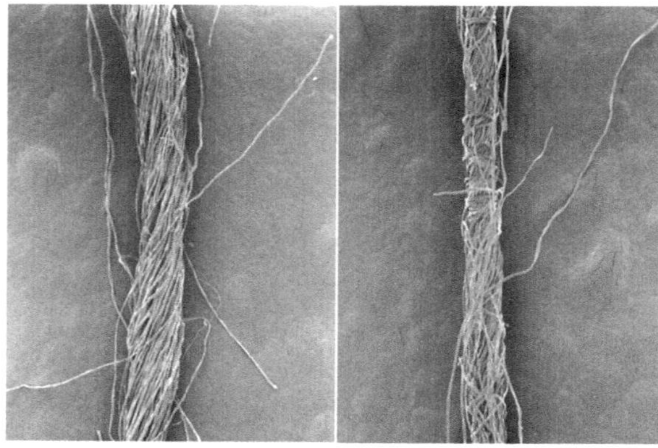

Fig. 10.1 Scanning electronic microscope images of a ring spun yarn (left) and an open-end or rotor spun yarn (right) (From Clapp 2001. Picture: M. J. Grimson)

the difference in yarn structure between the two. Figure 10.1 illustrates the different structure of the two yarn types. The most obvious difference between them is the presence of wrapper fibers on the rotor spun yarn. The sole force holding together ring spun yarn is lateral fiber-to-fiber friction, making length and fineness of fibers important because they affect the amount of contact among fibers. Fibers should be long enough to consistently overlap, and fineness maximizes the number of fibers that fit in a cross section of yarn. The structure of ring spun yarn also results in sensitivity to short fiber content (McCreight et al. 1997). Rotor spun yarn is held together by lateral friction and wrapper fibers, making fiber-to-fiber contact less crucial. However, the initial quality of raw fiber will affect yarn quality for either of these or any other spinning methods. Fiber traits such as length, strength, and surface properties are responsible for the strength and integrity of the yarn (Clapp 2001). A third and newer type of spinning technology, air jet vortex spinning, requires fibers with greater length uniformity to produce optimum yarns (Erdumlu et al. 2012). Ultimately, yarn quality is important because it will affect many properties of the fabric including but not limited to uniformity and dyeability (El Mogahzy 1999). It is the goal of the textile industry to produce high-quality end products while being financially and timely efficient.

Since the 1980s and the development of the high volume instrument (HVI) to objectively measure and grade physical properties of cotton fiber, cotton breeders and others in the industry have access to more complete data for determining fiber quality. Figure 10.2 provides a historical assessment of changes in US on-farm HVI fiber quality over a period of 30 years (data obtained from USDA-AMS, http://ww.ams.usda.gov). Changes in on-farm fiber quality data are due to a number of factors including genetics, management practices, harvesting and ginning practices, the growing environment, and interactions between genetics, management practices, and the growing environment. Overall, trends suggest increases for fiber length (a), fiber strength (b), and micronaire (c), while uniformity index (d) has been flat.

In this chapter, we will summarize the genetic improvement of cotton fiber quality. Efforts will be made to discuss the genetic improvement of cotton fiber quality in the context of plant breeding and biotechnology.

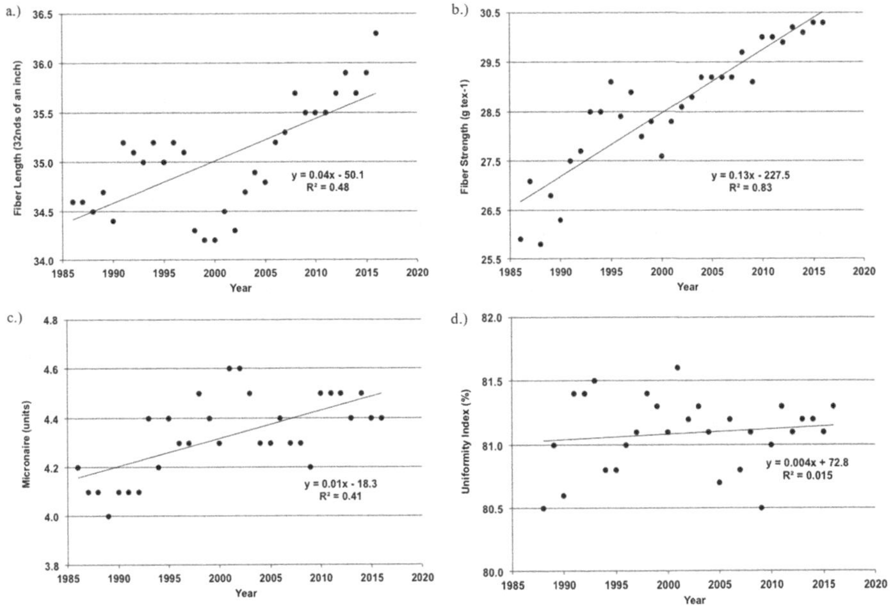

Fig. 10.2 On-farm US cotton fiber quality trends from 1986 to 2016 for (**a**) fiber length, (**b**) fiber strength, (**c**) micronaire, and (**d**) length uniformity

10.2 Improving Fiber Quality Through Plant Breeding

Genetic improvement efforts to improve cotton fiber quality based upon conventional plant breeding methods are dependent upon knowledge of the genetic architecture of fiber quality, extent of genotype × environment ($G \times E$) interactions on fiber quality, the relationship between fiber quality and other traits of interest (i.e., lint yield), and the extent of genetic diversity for fiber quality.

10.2.1 Gene Action and Inheritance

Fiber quality traits of interest to plant breeders are complex and inherited quantitatively (Meredith 1984). Thus, variation in cotton fiber quality for a defined population is the result of genetics, the environment, and interactions between genetics and the environment. Variation attributable to genetics can be further partitioned into three components representing different types of gene action: (1) additive genetic effects, (2) dominance genetic effects, and (3) gene × gene interactions (i.e., epistasis). Quantitative genetics theory enables the estimation of these components and their contributions to the total phenotypic (i.e., observed) variability for a given trait and provides plant breeders with the information necessary to design and implement effective genetic improvement programs.

Despite being quantitative in nature, classical genetic studies demonstrate that fiber quality traits, in particular fiber length, fiber strength, and fineness, are moderately to highly heritable, and the populations under study possessed adequate additive genetic variations to be amenable to selection (Al-Rawi and Kohel 1970; Baker and Verhalen 1973; Lee et al. 1967; May 1999; Verhalen and Murray 1969). Researchers also demonstrated that fiber quality traits were minimally influenced by $G \times E$ interactions (Abou-El-Fittouh et al. 1969; Al-Jibouri et al. 1958; Meredith and Bridge 1973; Miller et al. 1958). Although genetic estimates apply to specific populations and to the environment(s) in which they were evaluated (Holland et al. 2003), these inferences on the genetics of the aforementioned fiber quality traits typically still hold true.

Figure 10.3 provides a summary of recent studies investigating the gene action and inheritance of several major fiber quality parameters, including fiber length, uniformity index, fiber strength, micronaire, and fiber elongation, as measured by single instruments and HVI (Campbell et al. 2013, 2014, 2016; Cheatham et al. 2003; Jenkins et al. 2007, 2009, 2012; McCarty et al. 2004; Saha et al. 2006; Song et al. 2015; Wu et al. 2010; Yuan et al. 2005; Zeng et al. 2013). Most fiber properties exhibit equivalent proportions of additive and dominance genetic variance on average, with the exception of fiber strength, which exhibits twice as much additive variance compared to dominance variance. In contrast, dominance genetic effects are often the predominant source of genetic variation for fiber quality traits among populations derived from crosses between elite and exotic germplasm, suggesting that exotic germplasm may harbor novel alleles that interact with those accumulated within the elite germplasm pool in a nonadditive way (Campbell et al. 2014;

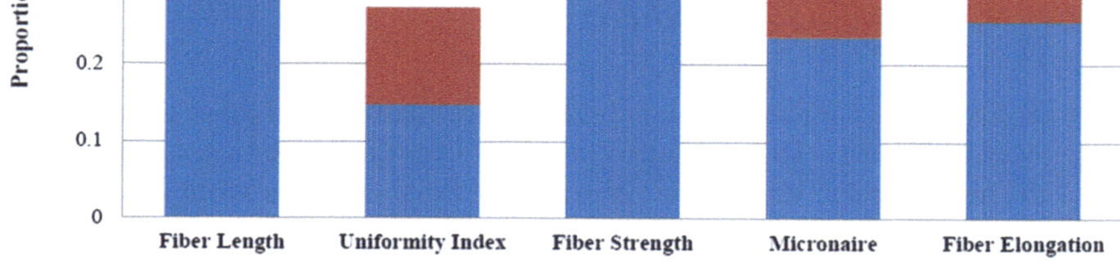

	Fiber Length	Uniformity Index	Fiber Strength	Micronaire	Fiber Elongation
V_A / V_P	0.30 (0.0 – 0.7)	0.15 (0.0 – 0.3)	0.35 (0.0 – 0.8)	0.24 (0.0 – 0.6)	0.26 (0.0 – 0.5)
V_D / V_P	0.27 (0.0 – 0.6)	0.13 (0.0 – 0.3)	0.17 (0.0 – 0.6)	0.22 (0.0 – 0.6)	0.21 (0.0 – 0.5)

Fig. 10.3 The percentage of the total variance (V_P) attributed to additive (V_A) and nonadditive (V_D) components of variance for fiber quality traits summarized from a range of published reports

Cheatham et al. 2003; McCarty et al. 2004). The nature of epistatic interactions has been difficult to measure using traditional quantitative genetic studies, whereas molecular genetics-based studies have allowed researchers to characterize interactions between specific loci contributing to quantitative variation (Tanksley 1993). For example, Shen et al. (2006) demonstrated that digenic interaction effects explained significant proportions of variation in fiber quality traits within a biparental population, ranging from 3.6 to 23.7% for fiber strength and uniformity index, respectively. Common inconsistency of marker-trait associations across environments and genetic backgrounds also suggests that epistatic genetic effects play an important role in the phenotypic expression of fiber quality traits (Fang 2015; Hugie et al. 2016). The development and use of resources such as genetic markers and genome sequence data will improve our ability to dissect gene action of fiber quality traits at the molecular level and facilitate continued germplasm improvement in cotton (Paterson et al. 2012; Hulse-Kemp et al. 2015; Li et al. 2015; Zhang et al. 2015).

Estimates of gene action enable the calculation of heritability, which ultimately determines the potential of a given trait for genetic improvement. Broad-sense heritability, H (i.e., the proportion of phenotypic variance attributed to genetic effects), estimates for fiber length, fiber strength, micronaire, and fiber elongation derived from recent studies are on average moderate to high, whereas the mean H for uniformity index is low (Campbell et al. 2013, 2014, 2016; Cheatham et al. 2003; Jenkins et al. 2007, 2009, 2012; McCarty et al. 2004; Saha et al. 2006; Song et al. 2015; Wu et al. 2010; Yuan et al. 2005; Zeng et al. 2013). Narrow-sense heritability, h^2 (i.e., the proportion of phenotypic variance attributed to additive genetic variation), estimates range widely from 0.0 to 0.8 but are moderate on average for all fiber quality traits, with the exception of uniformity index, demonstrating that sufficient additive genetic variation exists within *Gossypium* spp. to enable genetic improvement for most fiber quality traits through inbreeding and selection. Superior measures of fiber length distribution are available to breeders through the Advanced Fiber Information System (AFIS) technology, which measures properties of individual fibers (Bragg and Shofner 1993). However, Smith et al. (2010) reported comparable H values (0.3) for AFIS length distribution measures. Additionally, Kelly et al. (2013) observed similar responses to selection based on HVI-measured length uniformity compared to selection based on AFIS-measured fiber length distribution. Genetic progress from research focused on better understanding fiber length distributions is expected in the future.

10.2.2 Genotype × Environment Interactions

The relative contributions of genetics, the environment, and $G \times E$ interactions to the phenotypic expression of traits are important factors in the selection of superior genotypes as well as resource allocation within a breeding program. Figure 10.4 provides a summary of several studies investigating the role of genotypic, environmental, and $G \times E$ interaction effects in the expression of major fiber quality traits (Campbell and Jones 2005; Campbell et al. 2012; Blanche et al. 2006; Kerby et al. 2000; Meredith

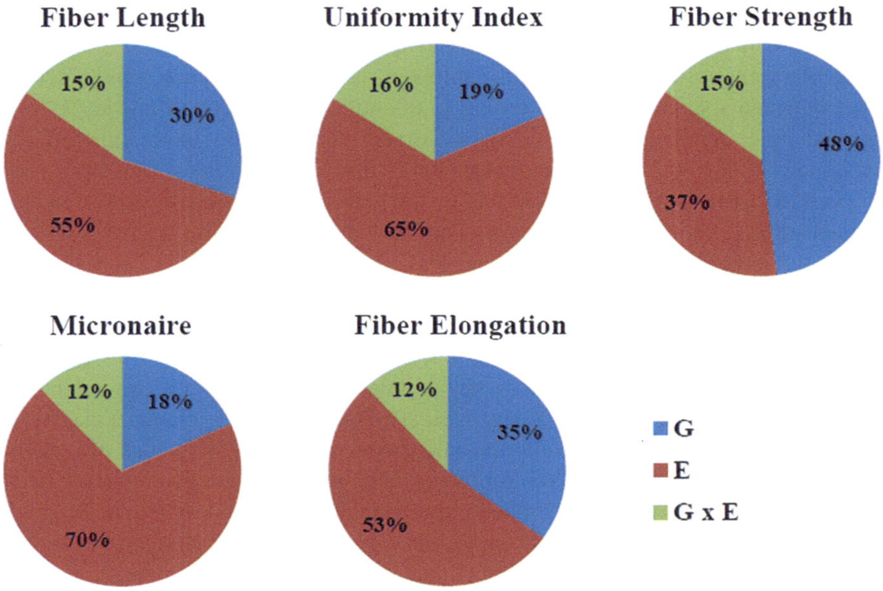

Fig. 10.4 A summary of the proportion of the total variance attributed to genetics (*G*), environment (*E*), and $G \times E$ interaction for fiber quality traits collected from a number of published reports

2003; Meredith et al. 2012). Environmental effects account for the majority of phenotypic variation observed for fiber quality traits, with the exception of fiber strength. The average proportion of phenotypic variability attributable to genetics ranges from 18% for micronaire to 48% for fiber strength. Genotypic effects are generally larger in comparison to $G \times E$ interaction effects for fiber length, fiber strength, and fiber elongation. Environmental effects and $G \times E$ interaction effects account for approximately 80% of the phenotypic variability in uniformity index and micronaire, both of which are considerably influenced by fiber maturity (Kelly et al. 2013).

There are two different types of $G \times E$ interactions: (1) $G \times E$ resulting from a change in magnitude regarding differences between genotypes (i.e., noncrossover interaction) and (2) $G \times E$ resulting from changes in the ranking of genotypes across environments (i.e., crossover interaction). The majority of $G \times E$ interactions identified for fiber quality traits are noncrossover interactions (Campbell et al. 2012; Campbell and Jones 2005; Meredith et al. 2012). Molecular genetics-based analyses have enabled further dissection of $G \times E$ interactions, such that the effects of individual loci contributing to the expression of fiber quality can be measured across different environments. Although $G \times E$ interactions for fiber quality traits generally account for less variation than main genotypic effects, molecular-based analyses suggest that the combination of genes contributing to the expression fiber quality varies depending on the environment, providing further evidence that fiber quality traits are controlled by a complex network of interacting genes (Lacape et al. 2010; Paterson et al. 2003).

10.2.3 Relationships Between Fiber Quality and Other Traits

Relationships between traits are also an important consideration in the design of selection schemes. Trait associations are most commonly measured as phenotypic (i.e., combined effects of genotypic, environment, and $G \times E$ interactions) or genetic (i.e., genotypic effects only) correlations between traits. Early genetic studies found that associations between fiber quality traits were generally favorable, whereas associations between fiber quality traits and lint yield components were typically unfavorable, making the simultaneous improvement of fiber quality and lint yield particularly challenging (Al-Jibouri et al. 1958; McCall et al. 1986; Meredith 1984; Miller et al. 1958; Miller and Rawlings 1967; Smith and Coyle 1997). Table 10.1 provides a summary of genetic correlations among major fiber quality and agronomic traits derived from a number of studies (Campbell et al. 2012; Fang et al. 2014; Lu and Myers 2011; McCarty et al. 2006; Zeng et al. 2007; Zeng and Meredith 2009). The relationships between fiber length, uniformity index, fiber strength, and micronaire are favorable on average. Longer fibers tend to have lower micronaire, owing to smaller fiber perimeter and a tendency for longer fibers to be less mature; longer, finer fibers result in superior bundle strength due to increased surface area and contact between fibers (May 2000; Price et al. 1999). Unfavorable associations between fiber quality traits and lint yield components still persist despite recent breeding efforts, in particular the negative correlations of fiber length and bundle strength with lint percent, boll number per plant, and lint yield (Table 10.1).

Genetic associations between traits can arise from pleiotropic effects (i.e., single gene influencing multiple traits) or linkage disequilibrium (i.e., nonrandom association of alleles at different loci). Several lines of evidence suggest that the negative relationships between fiber quality traits and lint yield are, at least in part, attributable to repulsion phase linkage. First, recurrent selection has been shown to reduce the negative correlation between lint yield and fiber strength (Culp et al. 1979; Miller and Rawlings 1967). Secondly, the magnitude of the negative association between fiber length and lint yield varies widely depending on genetic background (Constable and Bange 2007). Finally, multiple germplasm lines combining desirable fiber quality and lint yield have been identified (Campbell et al. 2012; Clement

Table 10.1 A summary of genotype correlations (and their standard errors) among fiber quality traits and agronomic traits summarized from a range of published reports

	Fiber length	Uniformity index	Fiber strength	Micronaire
Uniformity index	0.27 (0.70)	–	–	–
Fiber strength	0.46 (0.27)	0.67 (0.12)	–	–
Micronaire	−0.49 (0.10)	−0.10 (0.25)	−0.22 (0.22)	–
Lint percent	−0.34 (0.24)	−0.12 (0.31)	−0.43 (0.08)	0.47 (0.22)
Boll weight	0.30 (0.15)	0.06 (0.53)	−0.03 (0.18)	−0.19 (0.42)
Boll number	−0.36 (0.14)	−0.16 (0.46)	−0.36 (0.08)	0.54 (0.40)
Lint yield	−0.17 (0.30)	0.01 (–)	−0.35 (0.19)	0.29 (0.20)

et al. 2012; Culp and Green 1992). Breeding programs which employ one or more of the following strategies have been successful in identifying these rare recombinants: (1) the use of recurrent selection to break up large linkage blocks, (2) evaluation of adequate breeding population sizes, and (3) the use of selection thresholds or indices for fiber quality, such as the Q score, in early generations (Bourland and Jones 2012; Clement et al. 2015; Campbell and Myers 2015). The use of genetic markers in early generation selection for fiber quality also has the potential to mitigate the negative relationship between fiber quality and lint yield (Hugie et al. 2016).

10.2.4 Genetic Diversity

Useful genetic variability is necessary for plant improvement. Reports of moderate-to-high heritability estimates suggest that sufficient variation exists within upland cotton germplasm for continued genetic improvement of fiber quality traits through selection (Bowman et al. 1996; Campbell and Myers 2015). However, molecular-based studies indicate low levels of genetic diversity within cultivated upland cotton germplasm that could potentially limit progress toward improved fiber quality (Fang et al. 2013; Hinze et al. 2012; Lacape et al. 2007; Van Deynze et al. 2009). Nevertheless, much of genetic diversity present in *Gossypium* outside of the elite upland cotton gene pool remains largely underutilized, with trends moving toward increased genetic uniformity due to a reliance on the repeated use of closely related, elite germplasm as parents in the development of modern cultivars (Van Esbroeck and Bowman 1998).

Efforts to increase fiber quality in upland cotton have traditionally exploited interspecific hybridization. Two of the most influential sources of increased fiber quality in upland cotton are (1) Beasley's tri-species hybrid ((*G. arboreum* L. × *G. thurberi* Tod.) × *G. hirsutum*) and (2) *G. barbadense* L. and can still be found in the pedigrees of most modern upland cultivar/germplasm releases with exceptional fiber qualities (Bowman and Gutierrez 2003; Niles and Feaster 1984; Smith et al. 1999, 2009; Zhang et al. 2005). Although putative introgressions from the tri-species hybrid and *G. barbadense* into upland cotton are presumably responsible for increases in cotton fiber quality, only a few introgression products have been verified. A few studies reported the development of stable interspecific-derived germplasm lines developed through backcrossing or intermating followed by multiple generations of selfing (Cantrell and Davis 2000; Percy et al. 2006; Yu et al. 2013). More often than not, successful interspecific transfer of fiber quality into upland cotton has been limited due to genetic barriers between species resulting in hybrid breakdown, sterility, and elimination of donor alleles in later generations (Jiang et al. 2000; Stephens 1949; Zhang et al. 2014).

To complement traditional interspecific hybridization efforts, chromosome substitution lines (CSLs) and chromosome segment introgression lines (CSILs) have also been developed and used to examine interspecific genetic variation with the ultimate aim of facilitating introgression of fiber quality from *G. barbadense* (Saha et al. 2006; Wang et al. 2012). Other strategies being used to identify novel genetic diversity for fiber quality traits include the conversion of primitive upland cotton accessions into day-neutral germplasm lines that can be more readily utilized in breeding programs (McCarty et al. 2004) and mutation breeding (Bechere et al. 2013; Patel et al. 2014).

Negative agronomic effects due to linkage drag are perhaps a greater challenge regarding the exploitation of diverse germplasm resources (Young and Tanksley 1989). Molecular-assisted breeding methods will likely be a valuable strategy to efficiently incorporate novel genetic diversity for fiber quality into breeding programs. Quantitative trait loci (QTL) mapping is commonly used to identify beneficial alleles for fiber quality traits within upland germplasm and interspecific derived populations, and a number of QTL have been successfully introgressed into elite upland backgrounds through genetic marker-assisted selection (Zhang et al. 2003; Guo et al. 2005; Shen et al. 2011).

10.2.5 Relationship with Yarn Quality

Kuraparthy and Bowman (2013) examined fiber quality trends in upland cultivars grown in the USA from 1980 to 2012 and found that fiber strength increased by 25% and fiber length increased by 4%. However, the rate of genetic gain for fiber quality traits has been lower than that achieved for lint yield (Campbell et al. 2011). In addition to the aforementioned negative association with lint yield, slower genetic gain for fiber quality traits can also be attributed to a lack of market incentive. The lack of incentive is not only attributable to the current cotton market, which demands what one may consider average quality for the majority of final products, from sheeting to denim, but also to an incomplete understanding of the relationship between fiber properties and yarn quality. As a result, many fiber properties in upland cotton have neither been minimized nor maximized through selection (May 2000; Hugie et al. 2016). The establishment of optimum values for fiber properties across spinning technologies and end uses is likely to improve market incentive by providing values on which price support programs may be established (Bourland and Jones 2012; May 1999).

10.3 Improving Fiber Quality Through Biotechnology

Commercial genetically engineered (GE) traits in cotton up to 2017 include production input traits herbicide tolerance (HT) and heliothis insect resistance (IR). GE production traits can have an indirect impact on overall fiber quality by reducing weedy non-lint content or insect infestation fiber staining and contamination from honeydew. Output traits directly affecting fiber are not commercialized, but research using biotechnology approaches of gene silencing, recombinant DNA, and fiber-specific promoters has contributed to fiber development knowledge base. John and Stewart (1992) theorized genetic engineering in cotton would move beyond crop management to GE-assisted genetic modification for improving fiber and developing novel products. Fiber-specific promoters were developed as early as 1996 to facilitate utility of potential novel high-impact GE traits (John 1996).

John (1997) postulated in the year after commercial introduction of IR and HT cotton that stress resistance and fiber improvement would be the next target of genetic modification through genetic engineering techniques. He cites new developments in gene identification and transformation technologies as indications of advanced progress. Value capture realization from GE traits directly impacting fiber quality requires more than proof of concept from enhanced genetic technology and improved knowledge of fiber gene functions.

Fiber value and targets for cotton fiber improvement are historically defined by contributions of spinnable fiber properties to yarn quality (Ramey et al. 1977). Fiber length, length uniformity, strength, fineness, and maturity indirectly measured as micronaire are assigned premiums and discounts both in the US cotton classification system and within actual values in cotton fiber marketing. Fiber color and trash content are also valued and discounted, due to their impact on spinning efficiency beyond the technical contribution of fiber length, strength, and fineness to yarn quality. Heritable properties, especially fiber length and strength, have long been a target of improvement through classical breeding, including molecular marker development research. Cotton fiber properties are quantitatively inherited and complex, with ambiguous value assignment due to different fiber requirements for an array of textile products and effect of supply and demand of specific fiber profiles in time.

By the late 1990s, several genes for stress resistance and fiber modification were being tested in various laboratories (John 1997). Genes from bacteria, spinach, potato, and spider, among others, have been used to study proof of concept for improving cotton fiber quality through GE techniques. Biotechnology-based genetic engineering aimed at cotton fiber improvement generally falls into three categories: elucidating and validating specific genes directly contributing to fiber development, such as fiber initiation and elongation; improving the environment under which fiber develops and matures targeting stress genes associated with drought, salt, or cold tolerance; and impacting functionality of cotton fiber, such as changing fiber reactivity to alter downstream processing properties of cotton.

According to a recent review article on the status of genetic engineering in cotton (Chakravarthy et al. 2012), examples of genetic transformation by different gene transfer methods include directly targeting fiber development genes (Li et al. 2004, 2010; Zeng et al. 2012); improving fiber, quantity, and/or quality, via reduced abiotic stress (He et al. 2005; Pasapula et al. 2011; Xu et al. 2012; Jiang et al. 2012); and altering fiber functionality (John and Keller 1996).

10.3.1 Proof of Concept and Fiber-Enabling Transformation

The first reported study to demonstrate genes from other organisms can modify theproperties of cotton genetically engineered cotton with two of the three genes bacteria use to produce polyhydroxybutyrate (PHB), theoretically making cotton function similar to polyester (John and Keller 1996). The transgenic fibers were slightly better insulators than wild-type cotton fibers and could presumably better retain heat. In the course of this research, two genes, each corresponding to fiber mRNA E6, were isolated from cv. Coker 312 (*Gossypium hirsutum* L.) and Sea Island (*G. barbadense* L.). E6 is one of the predominant fiber-specific mRNAs present during early fiber development (John 1996). Transgenic plants containing antisense genes were generated with reduced E6 protein levels in the range of 60–98%. E6 was ultimately deemed not important to normal fiber development or structure since no phenotypic change was apparent in fiber from the transgenic plants. Rinehart et al. (1996) reported isolating and characterizing FbL2A gene expressed in Sea Island cotton fiber. Multiple FbL2A developmentally regulated genes in cotton are activated during late primary and early secondary wall synthesis stages. The FbL2A promoter was used to express PHB transgenes in cotton fiber (John and Keller 1996). These discoveries represent some of the first reported cotton fiber-specific promoter genes and encouraged subsequent GE-enabled cotton fiber genetic modification research. Kim et al. (2011) continued research at USDA-SRRC on promoter regions pertinent to cotton fiber improvement by studying GhCesA4 gene responsible for cellulose production in cotton fiber. Information is expected eventually to assist researchers develop strategies for enhancing cellulose production and contribute to improved cotton fiber quality.

10.3.2 Characterization of Genes Expressed in Fiber Development

GE-modified cotton fiber improvement initially targeted genes thought to directly impact development of spinnable fiber properties, fiber elongation, and secondary wall development. Ruan et al. (2003) studied the role of sucrose synthase (*Sus*) gene expression in cotton fiber in one of the first reports to identify genes critical to cotton fiber development. Suppression of *Sus* expression only in the maternal seed tissue inhibited fiber development without affecting embryo development and seed size. A transgenic strategy to overexpress *Sus* did not approach fiber quality improvements greater than can be achieved through classical breeding. Li et al. (2004) reported up to 15% increase in fiber length and strength by expressing *acsA* and *acsB* cellulose synthesis genes from *Acetobacter xylinum*. Li et al. (2009) transformed cotton with silkworm fibroin (*fbn*) gene and observed increases up to 28.4% in length and up to 8.3% in strength. Betancur et al. (2010) show CesA genes, believed to encode the catalytic subunit for cellulose synthase, can support secondary wall thickening of *Arabidopsis* shoot trichomes and thus may have potential to improve cotton fiber maturity. Another strategy for GE-based fiber improvement is transforming cotton with plant-derived expansin genes (Wilkins 2006; Zhu et al. 2006; Kamran et al. 2013). Most observed fiber improvements from transformed fiber-specific development genes to date fall within the normal range of fiber variation in *Gossypium*.

Characterization of genes that regulate cotton fiber development is expected to continue, and a large number of expressed sequence tags are publically available. New transformation technologies such as virus-induced gene silencing (Tuttle et al. 2015) have been used to more efficiently investigate gene function in cotton development. Qu et al. (2012) used tobacco rattle virus-induced gene silencing (TRV-VIGS) to report plants with suppressed KATANIN expression produced shorter fibers and silenced WRINKLED1 expression increased fiber length. Their results also indicate TRV-VIGS can be used for rapid functional analysis of genes involved in cotton fiber development.

Research ongoing at USDA-SRRC (Islam et al. 2016) identifies and validates a cluster associated with fiber quality traits, strength, length, uniformity, and short fiber content on chromosome A07. Candidate genes related to fiber quality are identified, particularly GhRBB1_A07 in upland cotton. Candidate genes were also identified specifically to improve fiber strength and yarn strength (Islam et al. 2016). Near isogenic lines (NILs) were developed that differ in individual fiber strength but are similar in other fiber properties. Comparisons between the lines show two signaling pathways possibly associated with high individual fiber strength, one ethylene and other phytohormonal pathways involved in fiber elongation, and the other receptor-like kinase signaling pathways involved in maintaining cell wall integrity. This strategy of isolating specific fiber properties with NILs helps clarify genotype/phenotype associations since many individual fiber properties can be correlated with each other. This type of research activity will be important as biotechnology-based cotton fiber improvement strategy moves beyond transgenic approaches to cisgenic approaches supported by advanced gene editing technologies.

10.3.3 Genetic Engineering Approaches to Improve Fiber Development

Another biotechnology-based strategy for cotton fiber improvement is targeting genes for transformation that improve the conditions under which fiber develops. Major components of cotton fiber development, fiber initiation, fiber elongation, and secondary cell wall development are all impacted by environmental conditions. Kim et al. (2013) observed fiber cell wall development to be associated with sensitivity to stress by comparing immature fiber (im) mutant and wild-type cotton using genomic analyses and biochemical assays. Their results could clarify fiber maturity-related molecular mechanisms regulated by response to environmental stress and facilitate cotton fiber quality improvement by manipulating expression of genes regulating secondary cell wall development. The im mutant has been used to study secondary wall development, and the Ligon lintless cotton mutant has been used to study fiber initiation and elongation (Wang et al. 2010). Abscisic acid (ABA) is a plant hormone involved in many biological processes including seed development, dormancy, germination, vegetative growth, and environmental stress responses, but with inhibitory effect on cotton fiber development. Gilbert et al. (2013) isolated a UDP-glycosyltransferase highly expressed in developing fiber of the mutant Ligon lintless-2 and postulate increased glycosylation of ABA during cotton fiber development can improve fiber quality. The UDP-glycosyltransferase UGT73C14 was deemed a good candidate for genetic manipulation to improve fiber quality.

Examples of transgenic approaches to improve resistance to stress and indirectly impact cotton fiber quality include superoxide dismutase (SOD) genes from tobacco, sucrose-phosphate synthase (SPS) genes from spinach, isopentenyl transferase (ipt) genes from *Agrobacterium tumefaciens*, and *Arabidopsis* AVP1. Superoxide dismutase genes were an early target for biotechnology research to improve stress tolerance in cotton. Payton et al. (1997) transformed cotton expressing a chimeric gene encoding a chloroplast-targeted Mn superoxide dismutase (SOD) from tobacco but demonstrated little difference in response to stress between the transformed and wild-type plants. Kim and Triplett (2008) reported previously identified SOD enzyme in cotton fiber secondary cell wall proven distinct from other SODs is present both in developing secondary cell walls and elongating primary walls of cotton fiber. The cell wall form of SOD is involved in primary and secondary wall development for controlling hydrogen peroxide levels, but manipulation of SOD genes through genetic engineering does not result in significantly improved fiber quality to date.

Cotton is a subtropical perennial plant, and fiber development is inhibited at low temperatures (Gipson and Ray 1969). Cellulose biosynthesis essentially ceases when temperature in the boll drops below 15 °C. Enhanced sucrose availability was hypothesized to improve cotton fiber quality under abiotic stress, so Haigler et al. (2007) produced transgenic cotton plants that overexpress spinach sucrose-phosphate synthase (SPS) because of its role in regulating sucrose synthesis in photosynthetic and heterotrophic tissues. Twelve independent transformation events were characterized for transgene stability, SPS expression, SPS protein production, and SPS activity in leaf and fiber. High expressing lines were characterized for carbon partitioning and fiber quality compared to wild-type and transgenic null controls. When exposed to cool night temperatures, transgenic line with highest SPS activity in leaf and fiber produced higher fiber micronaire and maturity ratio, indicating more cellulose synthesis under cool temperatures.

Targeting genes that might impart resistance to salinity has also shown some indirect impact on cotton fiber development. Pasapula et al. (2011) studied *Arabidopsis* gene AVP1 that encodes a vacuolar pyrophosphatase functioning as a proton pump on the vacuolar membrane. AVP1 overexpression previously enhanced *Arabidopsis*, tomato, and rice plant performance under salt and drought stress by sequestering ions and sugars in the vacuole and reducing water potential. AVP1 overexpression was also observed to stimulate auxin transport in the root system, allowing transgenic plants to absorb water more efficiently. Transgenic cotton plants similarly showed more vigorous growth and improved drought and salt stress in greenhouse conditions, along with higher fiber yield in drought-stressed field conditions. Overexpression of *Agrobacterium tumefaciens* isopentenyl transferase (*ipt*) gene under the control of *G. hirsutum* cysteine proteinase (*Ghcysp*) promoter increased the accumulation of cytokinins and chlorophyll but also resulted in improved fiber quality of transgenic cotton plants under saline conditions (Liu et al. 2012).

Cotton fiber develops under many conditions around the globe from year to year. Biotechnology-based approaches to either significantly change cotton fiber properties such as length and strength or improve conditions under which fiber develops to influence maturity variation have not risen to the level of novel product, or assurance of consistent quality, that might warrant translating basic research into commercial development. Biotechnology research targeting cotton fiber has increased knowledge on basic fiber development, genotype/phenotype associations, and impact of environmental stress and helps lay a foundation for future genetic improvement of cotton fiber. Interaction of cotton fiber development, plant physiology, and environmental stress is complex, beyond single-gene strategies, and practical value capture from biotechnology solutions may need to go beyond improvement of traditional fiber properties as defined by spinning performance.

10.3.4 Bioengineered Functional Fibers

The first attempt at foreign gene modification of cotton fiber did not target traditional spinning-related fiber properties such as length, strength, length uniformity, maturity, or elasticity ostensibly because those genes had not been identified but sought to synthesize PHB in cotton fiber to produce a new polymer, with different functions (John and Keller 1996). It was hoped to create a new product with the strength and wrinkle-fastness of polyester inside the polymer combined with cotton comfort and feel on the fiber surface, the best of both worlds. Fiber from transgenic cotton plants did contain PHB granules, which appeared to alter insulating properties, but the researchers concluded that a severalfold increase in PHB synthesis would be required for product applications, something that would likely have concomitant effects on plant and fiber development. Nevertheless, the concept demonstrated potential of biotechnology applications on cotton fiber research and suggested new generations of fibers could be developed through genetic engineering.

Breeding improvements have focused on improving cotton fiber physical properties, with less emphasis on chemical properties; value substitution for transgenic cotton has been at the farm gate, while altering chemical fiber properties could shift value to the textile industry. Altering reactivity of cotton fiber is hypothesized to affect chemical properties in a way that reduce finishes required in textile processing, as well as alleviate their disposal expense and impact (Trolinder-Wright 2007). Daniell (2008) describes an invention to genetically engineer cotton to increase fiber strength, water absorption, and dye binding with a gene encoding an elastic and plastic protein-based polymer. Another invention describes altering the reactivity of cotton fiber to improve colorfastness to decrease the volume of wastewater used during dyeing (De Block et al. 2013). Transgenic approach to modify fiber by inserting positively charged oligosaccharides or polysaccharides into the cell wall could also be applied to improve the reactivity of cotton fiber with flame retardant, water-oil-soil repellants, anti-wrinkling agents, and other fabric finishes.

Bioengineered fiber has been produced from cotton plants as proof of concept, and Ceylan et al. (2012) report on attempts to establish a test method to screen for presence and efficacy of transgenic fibers with improved reactivity via incorporation of positively charged nitrogen moieties. A method capable of accurately detecting increased nitrogen levels in bioengineered cotton fibers is proposed. Commercial prospects of bioengineered fiber may depend on addressing potential consequences of unintentional comingling with regular (not as reactive) cotton fiber. Fiber from plants transformed to increase their reactivity may differentiate in wet processing and potentially cause quality defects at the fabric stage. Conversely, unintentional comingling of fiber from cotton plants not transformed to enhance reactivity could potentially reduce value of high-impact output traits such as those in bioengineered fiber.

10.4 Future Prospects for Cotton Fiber Improvement

Future cotton fiber improvement will result from a combination of traditional breeding techniques, advances in harnessing our increased knowledge of the cotton genome, and cutting-edge biotechnological tools. Recent quantitative genetic studies suggest that sufficient phenotypic variation exists to continue small, incremental improvements in fiber quality realized through conventional breeding. Augmenting conventional breeding methods with new knowledge of the gene makeup and structural variation of the cotton genome potentially will pave the way for new methods of combining beneficial fiber quality alleles into new commercial cultivars.

In addition, extensive studies to better understand molecular basis of cotton fiber physiology help not only to elucidate underlying genetics of differences between fiber properties in varying genotypes but also identify the role specific polysaccharides play in final fiber characteristics. Lacape et al. (2012) compared the two cultivated allotetraploid species at two critical cotton fiber development points (10 days post anthesis, peak elongation; 22 days post anthesis, transition to secondary cell wall synthesis) to document the fiber transcriptomes using expressed sequence tag (EST) pyrosequencing. The new set of fiber ESTs with gene-based markers adds to resources for further fiber improvement in cotton. Rajasundaram et al. (2014) studied cotton fiber characteristics with glycan arrays using regression-based approaches, establishing a relationship between glycan measurements and phenotypic traits. Their characterization included more species, but looked at the mature fiber stage only. Recent publication of cotton genome sequences renews optimism that this type of basic research can translate to commercial cotton fiber improvement through breeding. In addition, increased understanding of cotton fiber development as demonstrated by Haigler et al. (2012) will continue to enhance our ability to improve cotton fiber quality.

Targeted genome editing with sequence-specific nucleases (i.e., zinc finger nucleases (ZFN) and transcription activator-like effector nucleases (TALEN) technologies) and platforms such as those based on the CRISPR-associated (Cas) endonuclease can facilitate fiber improvement strategies from transgenic to cisgenic approaches. The first studies describing genome editing in cotton have recently been reported (Li et al. 2017; Gao et al. 2017). Cotton fiber is a genetically complex trait, and genes needed for meaningful transformation could add up to an unmanageable number. Strategies to stack multiple traits in a single already exiting locus (D'Halluin et al. 2013), or target multiple genes through gene editing, may be necessary to use genetic engineering for significant cotton fiber improvement.

Genetic research to identify and characterize genes is highly valuable to the body of knowledge; commercial feasibility of targeting output traits such as cotton fiber with genetic engineering depends on translating that knowledge to crop improvement in a radically different manner than can be done by classical breeding. Molecular-based enabling technologies are intended to enhance classical breeding, by discerning genetic variation and its function—not discovering new variation, or creating genetic variation and changing its function. Genome editing applied to systematic identification of agriculturally relevant genes and trait improvement should be accompanied by strategies to significantly enhance genetic diversity, such as wide crosses and characterization of genetic resources including wild relatives.

References

Abou-El-Fittouh HA, Rawlings JO, Miller PA (1969) Genotype by environment interactions in cotton—their nature and related environmental variables. Crop Sci 9:377–381

Al-Jibouri HA, Miller PA, Robinson HF (1958) Genotypic and environmental variances and covariances in an upland cotton cross of interspecific origin. Agron J 50:633–636

Al-Rawi KM, Kohel RJ (1970) Gene action in the inheritance of fiber properties in intervarietal diallel crosses of upland cotton, *Gossypium hirsutum* L. Crop Sci 10:82–85

Baker JL, Verhalen LM (1973) The inheritance of several agronomic and fiber properties among selected lines of upland cotton, *Gossypium hirsutum* L. Crop Sci 13:444–450

Bechere E, Meredith WR, Boykin JC (2013) Registration of mutant population MD 15 M4 *Gossypium hirsutum* L. with enhanced fiber quality. J Plant Regist 7:216–219

Benedict CR, Kohel RJ, Lewis HL (1999) Cotton fiber quality. In: Smith CW, Cothren JT (eds) Cotton: origin, history, technology, production. Wiley, New York, pp 269–288

Betancur L, Singh B, Rapp RA, Wendel JF, Marks MD, Roberts AW, Haigler CH (2010) Phylogenetically distinct cellulose synthase genes support secondary wall thickening in Arabidopsis shoot trichomes and cotton fiber. J Integr Plant Biol 52:205–220

Blanche SB, Myers GO, Zumba JZ, Caldwell D, Hayes J (2006) Stability comparisons between conventional and near-isogenic transgenic cotton cultivars. J Cotton Sci 10:17–28

Bourland FM, Jones DC (2012) Registration of 'UA48' cotton cultivar. J Plant Regist 6:15–18

Bowman DT, Gutierrez OA (2003) Sources of fiber strength in the U.S. upland cotton crop from 1980 to 2000. J Cotton Sci 7:164–269

Bowman DT, May OL, Calhoun DS (1996) Genetic base of upland cotton cultivars released between 1970 and 1990. Crop Sci 36:577–581

Bragg CK, Shofner FM (1993) A rapid, direct measurement of short fiber content. Text Res J 63:171–176

Calhoun S, Bowman DT (1999) Techniques for development of new cultivars. In: Smith CW, Cothren JT (eds) Cotton: origin, history, technology, production. Wiley, New York, pp 361–414

Campbell BT, Chee PW, Lubbers EL, Bowman DT, Meredith WR, Johnson J, Fraser DE (2011) Genetic improvement of the Pee Dee cotton germplasm collection following seventy years of plant breeding. Crop Sci 51:955–968

Campbell BT, Chee PW, Lubbers E, Bowman DT, Meredith WR Jr, Johnson J, Fraser D, Bridges W, Jones DC (2012) Dissecting genotype × environment interactions and trait correlations present in the Pee Dee cotton germplasm collection following seventy years of plant breeding. Crop Sci 52:690–692

Campbell BT, Greene JK, Wu J, Jones DC (2014) Assessing the breeding potential of day neutral converted racestock germplasm in the Pee Dee cotton germplasm enhancement program. Euphytica 195:453–465

Campbell BT, Greene J, Wu J, Jones DC (2016) Genetic variation for agronomic and fiber quality traits in a population derived from high-quality cotton germplasm. Crop Sci 56:1689–1697

Campbell BT, Jones MA (2005) Assessment of genotype × environment interactions for yield and fiber quality in cotton performance trials. Euphytica 144:69–78

Campbell BT, Myers GO (2015) Quantitative genetics. In: Fang DD, Percy RG (eds) Cotton, Agronomy monographs 57, 2nd edn. ASA, CSSA, and SSSA, Madison, pp 187–204

Campbell BT, Weaver DB, Sharpe R, Wu J, Jones DC (2013) Breeding potential of elite Pee Dee germplasm in upland cotton breeding programs. Crop Sci 53:894–905

Cantrell RG, Davis DD (2000) Registration of NM24016, an interspecific-derived cotton genetic stock. Crop Sci 40:1208

Cantrell RG, Roberts CL, Waddell C (2000) Registration of 'Acala 1517-99' cotton. Crop Sci 40:1200–1201

Ceylan O, Van Landuyt L, Van der Schueren L, Hauben M, De Block M, De Clerck K (2012) Innovative screening of novel bioengineered cotton fibers containing olichitin. Text Res J 82:801–809

Chakravarthy VSK, Reddy TP, Reddy VD, Rao KV (2012) Current status of genetic engineering in cotton (*Gossypium hirsutum* L): an assessment. Crit Rev Biotechnol 34:144–160

Cheatham CL, Jenkins JN, McCarty JC Jr, Watson CE, Wu J (2003) Genetic variances and combining ability of crosses of American cultivars, Australian cultivars, and wild cottons. J Cotton Sci 7:16–22

Clapp DM (2001) Textile processing. In: Seagull R, Alspaugh P (eds) Cotton fiber development and processing: an illustrated overview. International Textile Center, Texas Tech University, Lubbock, pp 72–87

Clement JD, Constable GA, Stiller WN, Liu SM (2012) Negative associations still exist between yield and fibre quality in cotton breeding programs in Australia and USA. Field Crop Res 128:1–7

Clement JD, Constable GA, Stiller WN, Liu SM (2015) Early generation selection strategies for breeding better combinations of cotton yield and fibre quality. Field Crop Res 172:145–152

Constable GA, Bange MP (2007) Producing and preserving fiber quality: from the seed to the bale. In: Stephens M (ed) Proceedings of the World Cotton Research Conference. Lubbock, TX, September, pp 10–14

Culp TW, Green CC (1992) Performance of obsolete and current Pee Dee germplasm lines of cotton. Crop Sci 32:35–41
Culp TW, Harrell DC, Kerr T (1979) Some genetic implications in the transfer of high fiber strength genes to upland cotton. Crop Sci 19:481–484
D'Halluin K, Vanderstraeten C, Van Hulle J, Rosolowska J, Van Den Brande I, Pennewaert A, D'Hont K, Bossut M, Jantz D, Ruiter R, Broadvest J (2013) Targeted molecular trait stacking in cotton through targeted double-strand break induction. Plant Biotechnol J 11:933–941
Daniell H (2008) Genetic engineering of cotton to increase fiber strength, water absorption and dye binding. US Patent 7,329,802
De Block M, Meulewaeter F, Koch R, Essigmann B, Bayer CropScience NV (2013) Methods for altering the reactivity of plant cell walls. US Patent 8,507,755
El Mogahzy YE (1999) Fiber-to-fabric engineering: optimization of cotton fiber quality. In: Basra AS (ed) Cotton fibers. The Haworth Press, Binghamton, pp 339–376
Erdumlu N, Ozipek B, Oxenham W (2012) Vortex spinning technology. Text Prog 44(3-4):141
Fang DD (2015) Molecular breeding. In: Fang DD, Percy RG (eds) Cotton, Agronomy monographs 57, 2nd edn. ASA, CSSA, and SSSA, Madison, pp 255–288
Fang DD, Hinze LL, Percy RG, Li P, Deng D, Thyssen G (2013) A microsatellite-based genome-wide analysis of genetic diversity and linkage disequilibrium in upland cotton (*Gossypium hirsutum* L.) cultivars from major cotton-growing countries. Euphytica 191:391–401
Fang DD, Jenkins JN, Deng DD, McCarty JC, Li P, Wu J (2014) Quantitative trait loci analysis of fiber quality traits using a random-mated recombinant inbred population in upland cotton (*Gossypium hirsutum* L.). BMC Genomics 15:397–411
Gao W, Long WL, Tian X, Xu F, Singh J, Liu PK, Botella JR, Song C (2017) Genome editing in cotton with the CRISPR/Cas9 system. Front Plant Sci 8:1–12
Gilbert MK, Bland JM, Shockey JM, Cao H, Hinchliffe DJ, Fang DD, Naoumkina MA (2013) A transcript profiling approach reveals an abscisic acid specific glycosyltransferase (UGT73C14) induced in developing fiber of Ligon lintless-2 mutant of cotton (*Gossypium hirsutum* L.). PLoS One 8:e75268
Gipson JR, Ray LL (1969) Fiber elongation rates in five varieties of cotton (*Gossypium hirsutum* L.) as influenced by night temperature. Crop Sci 9:339–341
Guo WZ, Zhang TZ, Ding YZ, Zhu YC, Shen XL, Zhu XF (2005) Molecular marker assisted selection and pyramiding of two QLTs for fiber strength in upland cotton. Acta Genet Sin 32:1275–1285
Haigler CH, Betancur L, Stiff MR, Tuttle JR (2012) Cotton fiber: a powerful single-cell model for cell wall and cellulose research. Front Plant Sci 3:1–7
Haigler CH, Singh B, Zhang D, Hwang S, Wu C, Cai WX, Hozain M, Kang W, Kiedaisch B, Strauss RE, Hequet EF, Wyatt BG, Jividen GM, Holaday AS (2007) Transgenic cotton over-producing spinach sucrose phosphate synthase showed enhanced leaf sucrose synthesis and improved fiber quality under controlled environmental conditions. Plant Mol Biol 63:815–832
He C, Yan J, Shen G, Fu L, Holaday AS, Auld D, Blumwald E, Zhang H (2005) Expression of an Arabidopsis vacuolar sodium/proton antiporter gene in cotton improves photosynthetic performance under salt conditions and increases fiber yield in the field. Plant Cell Physiol 46:1848–1854
Hinze LL, Dever JK, Percy RG (2012) Molecular variation among and within improved cultivars in the U.S. cotton germplasm collection. Crop Sci 52:222–230
Holland JB, Nyquist WE, Cervantes-Martinez CT (2003) Estimating and interpreting heritability for plant breeding: an update. In: Jannink J (ed) Plant breeding reviews, vol 22. Wiley, Oxford, pp 9–112
Hugie KL, Fang DD, Smith CW, Li P, Hinze LL, Hague SS, Jones DC (2016) Utility assessment of published microsatellite markers for fiber length and bundle strength QTL in a cotton breeding program. Crop Sci 56:2983–2995
Hulse-Kemp AM, Lemm J, Plieske J, Ashrafi H, Buyyarapu R, Fang DD, Frelichowski J et al (2015) Development of a 63K SNP array for cotton and high-density mapping of intraspecific and interspecific populations of *Gossypium* spp. G3: Genes, Genomes, Genetics 5:1187–1209

Islam MS, Fang DD, Thyssen GN, Delhom CD, Liu Y, Kim HJ (2016) Comparative fiber property and transcriptome analyses reveal key genes potentially related to high fiber strength in cotton (*Gossypium hirsutum* L.) line MD52ne. BMC Plant Biol 16:36

Islam MS, Thyssen GN, Jenkins JN, Zeng L, Delhom CD, McCarty JC Jr, Deng DD, Hinchliffe DJ, Jones DC, Fang DD (2016) A MAGIC population-based genome-wide association study reveals functional association of GhRBB1_A07 gene with superior fiber quality in cotton. BMC Genomics 17:903

Jenkins JN, McCarty JC Jr, Wu J, Gutierrez OA (2009) Genetic variance components and genetic effects among eleven diverse upland cotton lines and their F2 hybrids. Euphytica 167:397–408

Jenkins JN, McCarty JC Jr, Wu J, Hayes R, Stelly D (2012) Genetic effects of nine *Gossypium barbadense* L. chromosome substitution lines in top crosses with five elite upland cotton *G. hirsutum* L. cultivars. Euphytica 187:161–173

Jenkins JN, McCarty JC Jr, Wu J, Saha S, Gutierrez O, Hayes R, Stelly DM (2007) Genetic effects of thirteen *Gossypium barbadense* L. chromosome substitution lines in topcrosses with upland cotton cultivars: II. Fiber quality traits. Crop Sci 47:561–572

Jiang CX, Chee PW, Draye X, Morell PL, Smith CW, Paterson AH (2000) Multilocus interactions restrict gene introgression in interspecific populations of polyploidy *Gossypium* (cotton). Evolution 54:798–814

Jiang Y, Guo W, Zhu H, Ruan YL, Zhang T (2012) Overexpression of GhSusA1 increases plant biomass and improved cotton fiber yield and quality. Plant Biotechnol J 10:301–312

John ME (1996) Structural characterization of genes corresponding to cotton fiber mRNA, E6: reduced E6 protein in transgenic plants by antisense gene. Plant Mol Biol 30:297–306

John ME (1997) Cotton crop improvement through genetic engineering. Crit Rev Biotechnol 17:185–208

John ME, Keller G (1996) Metabolic pathway engineering in cotton: biosynthesis of polyhydroxybutyrate in fiber cell. Proc Natl Acad Sci U S A 93:12678–12773

John ME, Stewart JM (1992) Genes for jeans: biotechnological advances in cotton. Trends Biotechnol 10:165–170

Kamran SB, Shahid AA, Rao AQ, Kiani MS, Ashraf MA, Dahab AA, Bakhsh A, Latif A, Khan MAU, Puspito AN, Aftab A, Bashir A, Husnain T (2013) Expression of Calotropis procera expansin gene CpEXPA3 enhances cotton fiber strength. Aust J Crop Sci 7:206–212

Kelly CM, Hequet EF, Dever JK (2013) Breeding for improved yarn quality: modifying fiber length distribution. Ind Crop Prod 42:386–396

Kerby TA, Burgess J, Bates M, Albers D, Lege K (2000) Partitioning variety and environmental contribution to variation in yield, plant growth, and fiber quality. In: Proceedings of the Beltwide Cotton Conference, San Antonio, TX, 4–8 Jan 2000. National Cotton Council of America, Memphis, TN, pp 528–532

Kim HJ, Murai N, Fang DD, Triplett BA (2011) Functional analysis of *Gossypium hirsutum* cellulose synthase catalytic subunit 4 promoter in transgenic Arabidopsis and cotton tissues. Plant Sci 180:323–332

Kim HJ, Tang Y, Moon HS, Delhom CD, Fang DD (2013) Functional analyses of cotton (*Gossypium hirsutum* L.) immature fiber (im) mutant reveal that fiber cell wall development is associated with sensitivity to stress. BMC Genomics 14:889

Kim H, Triplett BA (2008) Involvement of extracellular Cu/Zn superoxide dismutase in cotton fiber primary and secondary cell wall biosynthesis. Plant Signal Behav 3:1119–1121

Kuraparthy V, Bowman DT (2013) Gains in breeding upland cotton for fiber quality. J Cotton Sci 17:157–162

Lacape JM, Dessauw D, Rajab M, Noyer JL, Hau B (2007) Microsatellite diversity in tetrapoloid *Gossypium* germplasm: assembling a highly informative genotyping set of cotton SSRs. Mol Breed 19:45–48

Lacape JM, Claverie M, Vidal RO, Carazzolle MF, Guimarães-Pereira GA et al (2012) Deep sequencing reveals differences in the transcriptional landscapes of fibers from two cultivated species of cotton. PLoS One 7:e48855

Lacape JM, Llewellyn D, Jacobs J, Arioli T, Becker D, Calhoun S, Al-Ghazi Y et al (2010) Meta analysis of cotton fiber quality QTLs across diverse environments in *Gossypium hirsutum x G. barbadense* RIL population. BMC Plant Biol 10:132–155

Lee JA, Miller PA, Rawlings JO (1967) Interaction of combining ability effects with environments in diallel crosses of upland cotton (*Gossypium hirsutum* L.). Crop Sci 7:477–481

Li Y, Liu D, Tu L, Zhang X, Wang L, Zhu L, Tan J, Deng F (2010) Suppression of GhAGP4 gene expression repressed the initiation and elongation of cotton fiber. Plant Cell Rep 29:193–202

Li C, Unver T, Zhang BH (2017) A high-efficiency CRISPR/Cas9 system for targeted mutagenesis in cotton (*Gossypium hirsutum* L.). Sci Rep 7:43902

Li X, Wang D, Zhao X, Dutt Y (2004) Improvement of cotton fiber quality by transforming the acsA and acsB genes into *Gossypium hirsutum* L. by means of vacuum infiltration. Plant Cell Rep 22:691–697

Li FF, Wu SJ, Chen TZ, Zhang J, Wang HH, Guo W, Zhan TZ (2009) Agrobacterium-mediated co-transformation of multiple genes in upland cotton. Plant Cell Tissue Organ Cult 97:225–235

Li F et al (2015) Genome sequence of cultivated upland cotton (*Gossypium hirsutum* TM-1) provides insights into genome evolution. Nat Biotechnol 33:524–530

Liu YD, Yin ZJ, Yu JW, Li J, Wei HL, Han XL, Shen FF (2012) Improved salt tolerance and delayed leaf senescence in transgenic cotton expressing the Agrobacterium IPT gene. Biol Plant 56:237–246

Lu H, Myers GO (2011) Combining abilities and inheritance of yield components in influential upland cotton varieties. Aust J Crop Sci 5:384–390

May OL (1999) Genetic variation in fiber quality. In: Basra AS (ed) Cotton fibers: developmental biology, quality improvement, and textile processing. Food Products Press, New York, pp 183–229

May OL (2000) Quality improvement of upland cotton (*Gossypium hirsutum* L.). J Crop Prod 5:371–394

May OL, Jividen GM (1999) Genetic modification of cotton fiber properties as measured by single- and high-volume instruments. Crop Sci 39:328–333

McCall LL, Verhalen LM, McNew RW (1986) Multidirectional selection for fiber strength in upland cotton. Crop Sci 26:744–750

McCarty JC Jr, Jenkins JN, Wu J (2004) Primitive accession derived germplasm by cultivar crosses as sources for cotton improvement: I. Phenotypic values and variance components. Crop Sci 44:1226–1230

McCarty JC Jr, Wu J, Jenkins JN (2006) Genetic diversity for agronomic and fiber traits in day neutral accessions derived from primitive cotton germplasm. Euphytica 148:283–293

McCreight DJ, Feil RW, Booterbaugh JH, Backe EE (1997) Short staple yarn manufacturing. Carolina Academic, Durham

Meredith WR, Bridge RR (1973) Yield, yield component and fiber property variation of cotton (*Gossypium hirsutum* L.) within and among environments. Crop Sci 13:307–312

Meredith WR Jr (1984) Quantitative genetics. In: Kohel RJ, Lewis CF (eds) Cotton, Agronomy monographs 24. ASA-CSSA, SSSA, Madison, pp 131–150

Meredith WR Jr (2003) Thirty-six years of regional high quality variety tests. In: Proceedings of the Beltwide Cotton Conference, Nashville, TN, 6–10 Jan 2003. National Cotton Council of America, Memphis, TN, pp 2561–2563

Meredith WR Jr, Boykin DL, Bourland FM, Caldwell WD, Campbell BT, Gannaway JR, Glass K, Jones AP, Ma LM, Smith CW, Zhang J (2012) Genotype × environment interactions over seven years for yield, yield components, fiber quality, and gossypol traits in the regional high quality tests. J Cotton Sci 16:160–169

Meredith WR, Nokes WS (2011) Registration of MD 9ne and MD 25 high-fiber-quality germplasm lines of cotton. J Plant Regist 5:202–206

Meyer L (2017) Cotton and wool outlook. USDA-ERS Publication CWS-17c, pp 1–18

Miller PA, Rawlings JO (1967) Selection for increased lint yield and correlated responses in upland cotton. Crop Sci 7:637–640

Miller PA, Williams JC Jr, Robinson HR, Comstock RE (1958) Estimates of genotypic and environmental variances and covariances in upland cotton and their implications in selection. Agron J 50:126–131

Niles GA, Feaster CV (1984) Breeding. In: Kohel RJ, Lewis CF (eds) Cotton, Agronomy monograph 24. ASA-CSSA, SSSA, Madison, pp 201–231

Pasapula V, Shen G, Kuppa S, Paez-Valencia J, Mendoza M, Hou P, Chen J, Qiu X, Zhu L, Zhang X, Auld D, Blumwald E, Zhang H, Gaxiola R, Payton P (2011) Expression of an Arabidopsis vacuolar H+-pyrophosphatase gene (AVP1) in cotton improves drought- and salt tolerance and increases fiber yield in the field conditions. Plant Biotechnol J 9:88–89

Patel JD, Wright RJ, Auld D, Chandnani R, Goff VH, Ingles J, Pierce GJ, Torres MJ, Paterson AH (2014) Alleles conferring improved fiber quality from EMS mutagenesis of elite cotton genotypes. Theor Appl Genet 127:821–830

Paterson AH, Saranga Y, Menz M, Jiang CX, Wright RJ (2003) QTL analysis of genotype × environment interactions affecting cotton fiber quality. Theor Appl Genet 106:384–396

Paterson AH, Wendel JF, Gundlach H, Guo H, Jenkins J, Jin D, Llewellyn D et al (2012) Repeated polyploidization of *Gossypium* genomes and the evolution of spinnable cotton fibres. Nature 492:423–427

Payton P, Allen RD, Trolinder NL, Holaday SA (1997) Over-expression of chloroplast-targeted Mn superoxide dismutase in cotton (*Gossypium hirsutum* L., cv. Coker 312) does not alter the reduction of photosynthesis after short exposures to low temperature and high light intensity. Photosynth Res 52:233–244

Percy RG, Cantrell RG, Zhang J (2006) Genetic variation for agronomic and fiber properties in an introgressed recombinant inbred population of cotton. Crop Sci 46:1311–1317

Price JB, Calamari TA, Tao WY (1999) Yarn preparation, fabric formation, and finishing. In: Smith CW, Cothren JT (eds) Cotton origin, history, technology, and production. Wiley, New York, pp 751–792

Qu J, Ye J, Geng YF, Sun YW, Gao SQ, Zhang BP, Chen W, Chua NH (2012) Dissecting functions of KATANIN and WRINKLED1 in cotton fiber development by virus-induced gene silencing. Plant Physiol 160:738–748

Rajasundaram D, Runavot JL, Guo X, Willats WGT, Meulewaeter F, Selbig J (2014) Understanding the relationship between cotton fiber properties and non-cellulosic cell wall polysaccharides. PLoS One 9:e112168

Ramey HH, Lawson R, Worley S (1977) Relationship of cotton fiber properties to yarn tenacity. Text Res J 47:685–691

Rinehart JA, Peterson MW, John ME (1996) Tissue-specific and developmental regulation of cotton gene FbL2A: demonstration of promoter activity in transgenic plants. Plant Physiol 112:1331–1341

Ruan YL, Llewellyn DJ, Furbank RT (2003) Suppression of sucrose synthase gene expression represses cotton fiber cell initiation, elongation, and seed development. Plant Cell 15:952–964

Saha S, Jenkins JN, Wu J, McCarty JC, Gutierrez OA, Percy RG, Cantrell RG, Stelly DM (2006) Effects of chromosome-specific introgression in upland cotton on fiber and agronomic traits. Genetics 172:1927–1938

Shen X, Cao Z, Singh R, Lubbers EL, Xu P, Smith CW, Paterson AH, Chee PW (2011) Efficacy of qFL-chr1, a quantitative trait locus for fiber length in cotton (*Gossypium* spp.). Crop Sci 51:2005–2010

Shen X, Zhang T, Guo W, Zhu X, Zhang X (2006) Mapping fiber and yield QTLs with main, epistatic, and QTL x environment interaction effects in recombinant inbred lines of upland cotton. Crop Sci 46:61–66

Smith CW, Braden CA, Hequet EF (2010) Genetic analysis of fiber length uniformity in upland cotton. Crop Sci 50:567–573

Smith CW, Cantrell RG, Moser HS, Oakley SR (1999) History of cultivar development in the United States. In: Smith CW, Cothren JT (eds) Cotton origin, history, technology, and production. Wiley, New York, pp 99–172

Smith CW, Coyle GG (1997) Association of fiber quality parameters and within-boll yield components in upland cotton. Crop Sci 37:1775–1779

Smith CW, Hague S, Hequet E, Thaxton PS, Brown IN (2008) Development of extra-long staple upland cotton. Crop Sci 48:1823–1831

Smith CW, Hague S, Thaxton PS, Hequet E, Jones D (2009) Registration of eight extra-long staple upland cotton germplasm lines. J Plant Regist 3:81–85

Song M, Fan S, Pang C, Wei H, Liu J, Yu S (2015) Genetic analysis of fiber quality traits in short season cotton (*Gossypium hirsutum* L.). Euphytica 202:97–108

Stephens SG (1949) The cytogenetics of speciation in *Gossypium*. I. Selective elimination of the donor parent genotype in interspecific backcrosses. Genetics 34:627–637

Tanksley SD (1993) Mapping polygenes. Annu Rev Genet 27:205–233

Trolinder-Wright L (2007) Actions and plans for cotton biotech in the commercial sector – Bayer CropScience. In: Proceedings World Cotton Research Conference – 4 ICAC. https://www.icac.org/meetings/wcrc/wcrc4/presentations/data/papers/Paper2156.pdf

Tuttle JR, Haigler CH, Robertson D (2015) Virus-induced gene silencing of fiber-related genes in cotton. Methods Mol Biol 1287:219–234

Van Deynze A, Stoffel K, Lee M, Wilkins TA, Kozik A, Cantrell RG, Yu JZ, Kohel RJ, Stelly DM (2009) Sampling nucleotide diversity in cotton. BMC Plant Biol 9:125

Van Esbroeck G, Bowman DT (1998) Cotton germplasm diversity and its importance to cultivar development. J Cotton Sci 2:121–129

Verhalen LM, Murray JC (1969) A diallel analysis of several fiber property traits in upland cotton *Gossypium hirsutum* L. Crop Sci 9:311–315

Wang QQ, Liu F, Chen XS, Ma XJ, Zeng HQ et al (2010) Transcriptome profiling of early developing cotton fiber by deep-sequencing reveals significantly differential expression of genes in a fuzzless/lintless mutant. Genomics 96:369–376

Wang P, Zhu Y, Song X, Cao Z, Ding Y, Liu B, Zhu X, Wang S, Guo W, Zhang T (2012) Inheritance of long staple fiber quality traits of *Gossypium barbadense* in *G. hirsutum* background using CSILs. Theor Appl Genet 124:1415–1428

Wilkins TA, The Regents of the University of California (2006) Bioengineering cotton fiber properties. US Patent 7,060,874

Wu J, McCarty JC, Jenkins JN, Meredith WR (2010) Breeding potential of introgressions into upland cotton: genetic effects and heterosis. Plant Breed 129:526–532

Xu SM, Brill E, Llewellyn DJ, Furbank RT, Ruan YL (2012) Overexpression of a potato sucrose synthase gene in cotton accelerates leaf expansion, reduces seed abortion, and enhances fiber production. Mol Plant 5:430–441

Young ND, Tanksley SD (1989) RFLP analysis of the size of chromosomal segments retained around the TM-2 locus of tomato during backcross breeding. Theor Appl Genet 77:353–359

Yu J, Zhang K, Li S, Yu S, Zhai H, Wu M, Li X, Fan S, Song M, Yang D, Li Y, Zhang J (2013) Mapping quantitative trait loci for lint yield and fiber quality across environments in a *Gossypium hirsutum* x *Gossypium barbadense* backcross inbred line population. Theor Appl Genet 126:275–287

Yuan YL, Zhang TZ, Guo WZ, Pan JJ, Kohel RJ (2005) Diallel analysis of superior fiber quality properties in selected upland cottons. Acta Genet Sin 32:79–85

Zeng L, Meredith WR, Boykin DL, Taliercio E (2007) Evaluation of exotic germplasm population derived from multiple crosses among *Gossypium* tetraploid species. J Cotton Sci 11:118–127

Zeng L, Meredith WR Jr (2009) Associations among lint yield, yield components, and fiber properties in an introgressed population of cotton. Crop Sci 49:1647–1654

Zeng Q, Qin S, Song S, Zhang M, Xiao Y, Luo M, Ho L, Pei Y (2012) Molecular cloning and characterization of a cytokinin dehydrogenase gene from upland cotton (*Gossypium hirsutum* L.). Plant Mol Biol Report 30:1–9

Zeng L, Wu J, Bechere E (2013) Genetic effects and genetic values of fiber properties in F2 and F3 hybrids between germplasm lines and high yield cultivars. Euphytica 190:459–469

Zhang T, Hu Y, Jiang W, Fang L, Guan X, Chen J, Zhang J et al (2015) Sequencing of allotetraploid cotton (*Gossypium hirsutum* L. acc. TM-1) provides a resource for fiber improvement. Nat Biotechnol 33:531–537

Zhang JF, Lu Y, Adragna H, Hughs E (2005) Genetic improvement of New Mexico Acala cotton germplasm and their genetic diversity. Crop Sci 45:2363–2373

Zhang J, Percy RG, McCarty JC Jr (2014) Introgression genetics and breeding between Upland and Pima cotton: a review. Euphytica 198:1–12

Zhang T, Yuan Y, Yu J, Guo W, Kohel RJ (2003) Molecular tagging of a major QTL for fiber strength in upland cotton and its marker-assisted selection. Theor Appl Genet 106:262–268

Zhu SW, Gao P, Sun JS, Wang HH, Luo XM, Jiao MY, Wang ZY, Xia GX (2006) Genetic transformation of green-colored cotton. In Vitro Cell Dev Biol 42:439–444

Index

D. D. Fang (ed.), *Cotton Fiber: Physics, Chemistry and Biology*,
https://doi.org/10.1007/978-3-030-00871-0

The manufacturer's authorised representative in the EU is Springer Nature Customer Service Centre GmbH, Europaplatz 3, 69115 Heidelberg, Germany. If you have any concerns regarding our products, please contact ProductSafety@springernature.com

Printed and bound by CPI Group (UK) Ltd, Croydon, CR0 4YY
22/09/2026
02221062-0007